WASTE

THE BASICS

Waste: The Basics answers the questions: why are we facing a global waste crisis, and how can we effectively solve it? The book identifies the most common types of waste, its major producers, how we manage waste locally, regionally, and globally, and why this management is leading to more waste.

Written in a highly accessible style, the book begins with our own everyday mundane experiences of creating waste (those objects or materials we toss in the garbage or recycling bin) and shows how these practices are connected to a global system that manages waste ineffectively. Drawing on a wealth of historical documents and empirical research, Hird unpacks the complex relationship that waste has with global structures of capitalism, neoliberalism, international trade, poverty, racialized and gendered relations, and social injustice. Armed with the basic facts about our 'waste-maker' global society, the author concludes that only by understanding waste as a byproduct of how society is organized around extraction, production, and consumption may we solve our increasing waste crisis through refusal, reduction, reuse, and re-orienting our lives to fit planetary sustainability boundaries.

Waste is written for students and general readers interested in waste as a human health and environmental issue. It is for anyone curious about where objects really go once we put them in the trash or recycling bin.

Myra J. Hird is a Full Professor, elected Fellow of the Royal Society of Canada, and Queen's National Scholar in the School of Environmental Studies, Queen's University, Canada. Hird is Director of Waste Flows, an interdisciplinary research project focused on waste as a global scientific-technical and socio-ethical issue. Hird has published 13 books and over 90 articles and book chapters on a range of topics relating to science studies. Hird's twelfth book, written with Hillary Predko, is called *Extracting Reconciliation* and is published by Routledge. Hird represented Canada at the G7 Science Meeting on Plastic Pollution in Paris, France.

The Basics

The Basics is a highly successful series of accessible guidebooks which provide an overview of the fundamental principles of a subject area in a jargon-free and undaunting format.

Intended for students approaching a subject for the first time, the books both introduce the essentials of a subject and provide an ideal springboard for further study. With over 50 titles spanning subjects from Artificial Intelligence to Women's Studies, *The Basics* are an ideal starting point for students seeking to understand a subject area.

Each text comes with recommendations for further study and gradually introduces the complexities and nuances within a subject.

PHILOSOPHY OF LANGUAGE
ETHAN NOWAK

STOIC ETHICS
CHRISTOPHER GILL AND BRITTANY POLAT

POLITICAL PHILOSOPHY
BAS VAN DER VOSSEN

INTERVIEWING
MARK HOLTON

PHENOMENOLOGY (SECOND EDITION)
DAN ZAHAVI

WASTE
MYRA J. HIRD

Other titles in the series can be found at: https://www.routledge.com/The-Basics/book-series/B

Designed cover image: Getty Images

First published 2025
by Routledge
4 Park Square, Milton Park, Abingdon, Oxon OX14 4RN

and by Routledge
605 Third Avenue, New York, NY 10158

Routledge is an imprint of the Taylor & Francis Group, an informa business

British Library Cataloguing-in-Publication Data
A catalogue record for this book is available from the British Library

ISBN: 978-1-032-50428-5 (hbk)
ISBN: 978-1-032-50424-7 (pbk)
ISBN: 978-1-003-39842-4 (ebk)

DOI: 10.4324/9781003398424

Typeset in Bembo
by codeMantra

WASTE

THE BASICS

Myra J. Hird

LONDON AND NEW YORK

CONTENTS

FIGURES

ACKNOWLEDGEMENTS

I thank my editor, Grace Harrison, for her guidance and encouragement throughout this project. I thank my editorial assistant, Matthew Shobbrook, for his unwavering support, encouragement, and flexibility.

I thank the Social Sciences and Humanities Research Council of Canada for funding for some of the case studies described in this book. Parts of Chapters 3, 5, and 6 appear in Hird, M.J. (2022). *A Public Sociology of Waste*. Bristol University Press, and are used with permission. Part of Chapter 6 appears in Hird, M.J. and Dee, G. (2024). Consuming Less in Hird, M.J. (ed.), *Consuming the Environment*. Routledge, and is used with permission.

I thank Inis and Eshe, for all of it.

PREFACE

Why are we facing a global waste crisis? How can we effectively resolve it? *Waste: The Basics* answers these questions. *Waste* provides a critical overview of the major causes of, and solutions to, our global waste crisis. It identifies the main sources of waste, and how these major sources produce the most damaging types of waste to human health and the environment. The book also examines how waste is managed at global, national, and local levels. Finally, *Waste* explores ways in which we may reduce waste, as well as manage our legacy waste more effectively. The book's main argument is that the current ways in which we manage waste are increasing the global volume (and toxicity) of waste. To reduce waste, we need to understand the complex connection between waste and our global society.

DEFINING WASTE

OVERVIEW

Chapter 1 introduces waste as a global environmental and human health crisis. It details why waste is a global human health problem, and how waste impacts the environment through climate change and pollution. The chapter defines the diverse types of waste that our global society produces, including: municipal solid and liquid waste; electronics waste; plastics waste; manufacturing waste; agricultural waste; radioactive waste; industrial, commercial, and institutional waste; and fossil fuels waste.

INTRODUCTION

Waste is simple to define. It is anything and everything we throw away. Waste, garbage, and/or discards are what we put in our trash cans, our bins – or, if we are less courteous, it is what we toss out the car window while driving down the road, or drop from our hands while walking down the street. Waste is what we pay our local governments to take away from our homes, to clean off our streets, and to take somewhere else. We may not like paying increasing taxes for this service, but the alternative – streets crammed with smelly flimsy bags of garbage that quickly invite critters – is worse. As anyone who has lived in New York, Montréal, or any other major

DOI: 10.4324/9781003398424-1

city during a garbage collection strike knows, the alternative is as distasteful as it is a threat to our health and wellbeing.

On closer reflection, waste's definition complexifies. When waste is diverted from disposal in favor of recycling, it ceases to be labeled waste, and becomes understood as a resource instead; for instance when rubber tires are recycled into road pavement. The same is the case when waste is burned to produce energy (for instance, when energy-from-waste facilities produce electricity that heats buildings). In the same way, electronics that people are no longer using in their homes are called 'hibernating stocks' rather than waste because they have not (yet) been physically discarded. More and more, politicians and industry are focused on ways to involve waste in the Circular Economy, such that waste may be used to produce more products. As such, what would otherwise be defined as waste (and therefore disposed of) is increasingly defined as a resource (see Chapter 2).

According to the United Nations and the World Bank, we are globally producing about 2.01 billion tons of municipal solid waste (MSW) (Kaza et al. 2018; UNEP 2024). Global waste generation is anticipated to increase to 3.8 billion tons by 2050. High income countries generate most of the world's waste: sixteen percent of the world's population generates 683 million tons of waste. As a country's income increases, so does their waste footprint. Whereas in low- and middle-income countries, food and green waste make up more than fifty percent of waste, high-income countries produce about the same amount of organic waste, but it only constitutes about thirty-two percent of the total waste because high-income countries produce more packaging and other non-organic waste. Recycling rates are increasing, with high-income countries recycling up to fifty percent of some materials (such as metal, cardboard, and glass). But waste generation, in these same countries, is also increasing.

In at least a third of the world, waste is not collected or disposed of in an environmentally safe way. Whereas high- and middle-income countries have waste collection and disposal programs and infrastructure, in low-income countries only about half of waste in cities is collected, and less than thirty percent of waste is collected in rural areas. Globally, only about thirty-seven percent of waste is disposed of in landfills; thirty-three percent is openly dumped; nineteen percent is recycled or composted; and about eleven percent is incinerated (Kaza et al. 2018).

WHY WASTE IS A PROBLEM

Waste is a problem because it is a significant contributor to climate change. In 2015, the World Meteorological Organization reported that the planet's temperature was one degree Celsius higher than the pre-industrial average. The United Nations warns that a two degree Celsius increase will lead to catastrophic climate change. If we do not reduce our greenhouse gas (GHG) emissions by half in less than ten years, we will exceed a 1.5 degree Celsius increase (IPCC 2018), with significant negative human health and environmental consequences.

In 2012, the United Nations declared waste to be a global environmental and human health problem (UN 2012). Through GHG emissions, we are affecting climate change: while carbon dioxide contributes about seventy-six percent of total GHG emissions, methane has twenty-one times the warming potential of carbon dioxide, and nitrous oxide has 310 times the warming potential (Connecticut Department of Energy and Environmental Protection 2020). Waste decomposition creates both methane and nitrous oxide. Plastics alone – because they are made from oil and gas – create significant carbon emissions. And transporting waste and recycling (see Chapter 4), is a very significant contributor to GHG emissions.

Waste is a problem because it creates pollution. Air pollution is created when harmful gases and chemicals are released into the air. Very small particles can enter our respiratory system when we breathe. Most air pollution is created through human actions such as burning fossil fuels, manufacturing products, producing food, and other industry emissions. Waste itself can create air pollution when gases and chemicals evaporate from it. Open dumps certainly create air pollution but so can landfills as microorganisms create methane, which is a greenhouse gas. Open dumps alone contribute eleven percent of the globe's particulate matter pollution (Hoesly et al. 2018). Incinerators and energy-from-waste facilities also create air pollution as waste is burned – and these gases and chemicals must be carefully collected using filter systems. In short, waste is a problem because – whether we dump, bury, or burn it – it contains harmful gases and chemicals. And these substances can leak into the surrounding ground, water, and air (Hird 2021, 2022).

Litter may be unintentional or intentional and can remain on the ground or in water for long periods before it somewhat or entirely biodegrades. Biodegradation means that the litter breaks into smaller pieces, and leaks into land, water, and air. We discard about 4.5 trillion cigarette butts alone each year (Novotny and Zhao 1999). A significant proportion of litter is made up of packaging, and especially fast-food packaging. Some litter, such as plastics like polystyrene, can remain in the environment for several decades or centuries before degrading into carbon dioxide and organic carbon (Ward et al. 2019). About twenty percent of the globe's human-made litter travels through stormwater, where it moves through streams, rivers, and waterways such as oceans. Ocean litter either washes up on beaches or it collects in what are known as ocean gyres such as the Great Pacific Garbage Patch (National Geographic 2024). Most (about eighty percent or more) of ocean litter comes from land (not fishing boats). Degraded litter cannot be recycled and is either openly dumped or goes to landfills. Wildlife such as birds, fish, and other animals can suffocate on litter that they eat or be otherwise strangled by litter such as plastics can-holders. Broken glass can cut animals and litter can poison wildlife when ingested.

Waste is a problem because we are increasing our extraction of the world's remaining raw materials, which is creating more waste. According to the 2023 Circularity Gap Report, about 100 billion tons of raw materials enter our manufacturing system each year (Circle Economy 2023). We are creating more waste because we are extracting, producing, and consuming more things. Many of these things are designed to be single-use or short-use such as plastics service ware (knives, forks, spoons, straws) and packaging. This increasing volume of waste – which is also often increasingly toxic as we create new synthetic materials – must be dealt with.

Waste is a problem because there is no easy fix. Open dumping is a human-health and environmental hazard. Landfilling waste means burying waste in the ground forever. Modern engineered landfills are designed to contain waste safely for relatively short periods of time (up to thirty-five years or so). And landfills create their own waste, known as leachate, which must also be contained. When leachate seeps out of landfills, it can create serious human health and environmental problems. For instance, high-profile cases such as Love Canal in Niagara Falls, New York, speak to the catastrophic consequences of open dumping or badly engineered landfills. Love Canal was the subject of a twenty-one year Superfund clean-up of toxic chemicals buried in a landfill. This toxic waste was covered over, and through a series of property sales ended up as the site of a residential community, which included schools, playgrounds, and so on (Smith 1982). The surrounding land and water were badly contaminated, and children and women were particularly affected by the contamination, with increased levels of particular cancers, neurological problems, and birth defects.

Incinerators and energy-from-waste facilities create their own waste as well. These facilities produce extremely toxic fly ash that must be disposed of in specially designed hazardous waste landfills (Hird 2021). Composting certainly reduces organic waste, but the energy and environmental resources (such as water, land, and considerable chemicals and nutrients in the case of agriculture) means that reducing the production of organic waste is far better for the environment than composting once organic waste has been produced.

And seventy-eight years after the bombing of Hiroshima and Nagasaki, the many problems associated with the long-term disposition of high-level nuclear wastes have remained unsolved. To date, several nations have geological repositories in development and testing (e.g. Finland, Sweden, France, Switzerland), and other nations are in the process of locating potential sites (e.g. Canada, United States, China, Holland) (Braden and Macfarlane 2023; IAEA 2022). Globally, we are generating about 12,000 tons of high-level nuclear waste every year (World Nuclear Association 2022a). And high-level nuclear waste has a half-life of tens of thousands of years. Both nuclear waste repositories and landfills consume enormous amounts of energy derived from fossil fuels to sort, treat, store, and transport waste (Chong and Hermreck 2010).

In sum, when we burn or bury waste, it doesn't really disappear. Waste is a global problem because it has harmful effects on human health and the environment. The many types of waste compound the risks to health and the environment through diverse toxicities and contaminations. The primary way that governments pay for waste to be managed is through taxes. As such, increasing waste (and its management) means increasing taxes.

TYPES OF WASTE

Countries and organizations (such as the United Nations, the World Bank, and Greenpeace) define several different general types of waste. What is included in these definitions may vary, depending on the region or country. For instance, France classifies radioactive waste by its activity and half-life, where activity refers to the amount of radiation emitted by the radioactive elements (radionuclides) contained in the waste, and half-life refers to the amount of time needed to half the radioactive activity (Garcier 2014). By contrast, Japan defines radioactive waste by its production chain, and Germany defines radioactive waste according to the heat it releases (ANDRA n.d.). Thus, the definitions provided below are necessarily broad, and may differ according to country.

MUNICIPAL SOLID WASTE

The waste that we are most familiar with is all the things we throw away in our daily lives, including a diverse range of objects such as plastic laundry soap containers, food scraps, polystyrene packaging, pet feces, pizza boxes, furniture, batteries, yard cuttings, and much more. All of these things are grouped within the category of MSW. In the popular press, MSW is often referred to as consumer waste or post-consumption waste. MSW also mixes non-hazardous and hazardous materials together. Hazardous waste is defined as any material that poses a risk to human health or the environment. A material may be hazardous because it easily ignites (and therefore poses a fire threat), is reactive (meaning that it may combine with other materials to cause a fire, explosion, or other risk), corrosive (in which case it could damage a landfill liner, causing waste to leak out of the landfill and into the environment), and/or toxic, (meaning that it may cause harm to human and non-human bodies, plants, and so on). While hazardous waste is usually placed in the category of industrial waste (called household hazardous and special wastes, or HHSW) because MSW and industrial waste come from different sources, MSW may actually have higher levels of toxins than industrial waste, depending on what the MSW dump or landfill includes (Wynne 1987).

A significant amount of MSW waste consists of packaging, and over eighty percent of packaging is plastics (United States Environmental Protection Agency 2022). Some packaging is used in order to protect the product while it is being shipped from the manufacturer to the retailer and on to the consumer. This transportation creates its own waste, including carbon emissions. Retailers use some packaging as an anti-theft technique: small products such as batteries are sold in packaging that is bigger than the batteries themselves. In many instances, the smaller the product, the bigger the package. But packaging is mainly designed to entice the consumer to buy the product. Companies use packaging to advertise the features of their products, including now advertising the features of the packaging itself. Within the food industry, for instance, this includes companies who advertise that their products are eco-friendly (such as plant-based rather than animal-based food) and that the packaging is eco-friendly (such as packaging made from recycled rather than new materials). As Chapter 2 details, some types of packaging are used because it is cheaper for the producer. This is the case, for instance, with plastics soda bottles, which are cheaper for the producer than glass bottles. Consumers are almost entirely responsible for disposing of products and their product packaging.

MUNICIPAL LIQUID WASTE

Along with MSW, individuals and households produce municipal liquid waste (MLW), which consists of all of the things that we dispose of via toilets, sinks, showers, bathtubs, washing machines, and so on. As wastewater management facilities readily attest, this comprises much more than human urine and feces: hygiene products – such as tampons, ear cleaners, toenail clippings, and hair – as well as condoms, toilet paper, fats, oils, grease, motor oil, solvents, gasoline, and a surprisingly diverse range of other objects are flushed down toilets. In many homes around the world, sanitary sewers (i.e. sewers that engineers have designed to carry wastewater to a treatment plant) are connected to kitchen and bathroom sinks, dishwashers, floor drains, and toilets. Sewers carry what is known as blackwater, which means water that carries human fecal matter and/or human urine. Sewers also carry greywater, which means water used from sanitary appliances such as clothes washing machines. Storm sewers, by contrast, transport untreated rain water directly to rivers or oceans (Mester et al. 2022). Sanitary sewers back up when objects clog the sewer lines. Wastewater treatment plants are designed to filter out objects, and then use chemical processes (such as oxidation) and/or ultraviolet radiation to kill microorganisms and other contaminants that are harmful to human health and/or the environment. The separation of objects from the water is typically done using a process of sedimentation (material sinks to the bottom of a water container and is collected) and/or flotation (material floats to the top of a water container and is collected), which culminates in a kind of sludge that is then disposed of in landfills (Center for Disease Control 2022). Other microorganisms are used to consume biodegradable organic contaminants (such as soaps and detergents), which are then separated from the water. The goal of any wastewater treatment plant is to sufficiently separate the waste from the water such that the latter may be returned to the water cycle (i.e. rivers, oceans, seas, but also drinking water) according to the regulations of that region that specify acceptable impact on human health and the environment.

MANUFACTURING WASTE

Manufacturing waste consists of any waste produced in the process of producing products: it is all of the unwanted raw material that is left over in the manufacturing process. Manufacturing waste may be solid, semi-solid, liquid or gaseous in form, and may be hazardous or non-hazardous. Manufacturing products produces waste, which may increase when there are product defects, by overproducing a particular product and/or by overprocessing a product. Globally, not all manufacturing industries are required to report the amount or kinds of waste they produce, making it difficult to estimate how much (and how toxic) manufacturing waste is.

Manufacturing waste includes all of the pollution that industries produce in the manufacturing of products. According to Environment Protection, manufacturing and construction produces about 6.3 billion tons of GHG emissions per year, and the fashion manufacturing industry creates 2.1 billion tons of GHG emissions per year (in Howell 2022). In Europe, the biggest manufacturing industries make motor vehicles and other transport equipment, basic and fabricated metals, and food, beverages, and tobacco products. The European Environment Agency's report *Costs of Air Pollution from European Industrial Facilities 2008–2017* estimates that a small number of industries within the UK, Germany, Poland, Spain, and Italy are producing fifty percent of industrial air pollution (Schecht et al. 2020). People living closer to manufacturing sites typically experience more air pollution.

An increasing number of products contain per- and poly-fluorinated alkyl substances (PFAS) or what are more commonly known as 'forever chemicals'. Manufacturers use PFAS chemicals in a wide range of products, from food packaging and textiles to cosmetics and electronics (Straková, Schneider, and Cingotti 2021; Straková, Brosché, and Brabcová 2023).

INDUSTRIAL, COMMERCIAL, AND INSTITUTIONAL WASTE

Industrial, commercial, and institutional (ICI) waste includes all of the waste produced by industries other than oil, gas, and mining. ICI waste includes all construction waste, retail waste, and wastes produced by institutions such as hospitals, schools, retirement homes, and so on. Some of this waste resembles MSW and MLW.

Other waste is more specific to the source. The World Health Organization (2018), for instance, notes that hospitals and other health care institutions produce: infectious waste (waste contaminated with bodily fluids such as blood and laboratory wastes contaminated with infectious agents); pathological waste (human tissues, organs, body parts, and animal carcasses); sharps waste (syringes, needles, disposable scalpels etc.); chemical waste (solvents, reagents, disinfectants, batteries, medical devices); pharmaceutical waste (unused, contaminated, or expired drugs and vaccines); cytotoxic waste (containing highly hazardous genotoxic substances, such as cancer treatment drugs); radioactive waste (containing radionuclides such as radioactive diagnostic material or radiotherapeutic materials), and non-hazardous general waste (waste that does not pose any biological, chemical, radioactive, or physical hazard to humans). About eighty-five percent of health care waste is composed of general non-hazardous waste, while fifteen percent comes from one or more of the hazardous categories.

Construction waste refers to all of the waste created during the construction, renovation, and demolition of buildings, roads, and bridges. Much of this waste is hazardous (such as batteries, fluorescent lights, electrical equipment) and must be disposed of in engineered landfills. Solid waste from construction and demolition accounts for twenty-three percent of total waste production in the United States (Bureau of Transportation Statistics 2020). There is considerable variability to the category of construction waste depending on the construction site.

Retail waste includes all of the waste that retailers (from 'big box' to 'mom-and-pop' stores) create. A very significant part of retail waste is generated from shipping retail products purchased online around the globe (Amazon, Etsy, and so on). United States retailers alone spent 1.5 trillion dollars on shipping and warehousing retail products in 2017 (Martinez 2020). Returns alone create over fifteen million metric tons of carbon dioxide emissions annually. And much of these returned goods end up in landfills because returned products are damaged in transit or it costs less to landfill products than it does to put them back on shelves for consumers to purchase. Optoro (2022) estimates that as many as 9.5 billion pounds of returned consumer products were disposed of in landfills in 2022: that's equivalent to 10,500 fully loaded Boeing 747s.

ELECTRONICS WASTE

Electronics waste (sometimes called waste from electrical and electronic equipment – WEEE; or end-of-life electronics – EOL; also called e-waste) includes a wide range of discarded electrical products and electronic devices, such as large and small household appliances, medical devices, automatic dispensers, smart phones, toys, desktop and laptop computers, and vehicles such as airplanes and cars – both of which are increasingly described as 'computers on wheels' (McKendrik 2022). Electronic waste is rapidly increasing as constantly newer versions of products (smart phones and watches, laptops, cars) are marketed.

Electronics are also subject to planned obsolescence, which means that companies intentionally design products to have short lives. Planned obsolescence happens, for instance, when companies such as Apple change the port standard on newer products so that consumers must buy new ports or when computers are designed to be unable to upload newer operating systems, which pressures consumers to buy new computers. The combination of rapidly advancing technology, the constant creation of new and revised products, planned obsolescence, and other factors has led to a sharp increase in electronic waste. Indeed, electronic waste is one of the fastest-growing global waste streams, increasing sixty percent between 2010 and 2019 (Statista n.d.; 2023). If current upward trends continue, we are on track to produce almost seventy-five million metric tons of e-waste by 2030. Northern European countries produce the most e-waste, followed by Australia and New Zealand, with Canada and the United States not far behind, and countries in Eastern Europe, Asia, and Africa producing significantly less electronic waste (Carrington 2020). In 2019, people in countries like France, the UK, and Germany produced 22.4 kg of e-waste per capita, compared to 2.5 kg of e-waste per capita in Chad, West Africa, and Zimbabwe.

Electronic waste is very valuable. According to The United Nations *Global E-waste Report*, e-waste contains over ten billion US dollars' worth of raw materials, including palladium, copper, platinum, gold, and other precious metals (in Carrington 2020). This same report found that e-waste is increasing faster than the world's human population, while only seventeen percent was recycled in 2019. E-waste is a global problem because the components contain harmful materials such as brominated flame retardants, cadmium, arsenic beryllium, and lead, which – if not disposed of safely, may lead to significant health risks to both the people who are disassembling electronics and the communities in which electronic-disassembly takes place. In countries in Asia and Africa, much e-waste disassembly takes place informally, where regulations and safety standards are minimal or non-existent, and electronics are burned to recover valuable materials, creating toxic fumes that humans inhale (see Chapter 5). Moreover, many wealthy countries export their e-waste to developing countries, exporting the health and environmental pollution problems with them (see Chapter 5).

AGRICULTURAL WASTE

Agricultural waste consists of any materials that are discarded in agricultural processes. Livestock manure constitutes a significant portion of agricultural waste. Globally, animals (primarily chickens, cattle, and sheep) produce about four times the amount of feces as humans. In 2014, the total mass of feces was 3.9×10^{12} or 3.9,000,000,000,000 kg, and is anticipated to increase to at least 4.6×10^{12} by 2030 as the world's human population grows and there is a rising demand for food (Berendes et al. 2018). Globally, people consume about seventy-five pounds of meat per person per year, constituting a more than doubling of meat consumption since 1990 (OECD 2015). Europeans alone consume about 165 grams of meat per person per day (Our World in Data n.d. a). Meat production requires a disproportionate amount of crop production: one kilogram of beef requires about twenty-five kilograms of crops (Our World in Data n.d. b)

Manure is a significant contributor to climate change, creating 11.1 percent to 19.6 percent of the world's GHG emissions (Blaustein-Rejto 2023). Besides carbon dioxide, manure releases other gases, including methane, nitrogen, ammonia, phosphorus, and nitrous oxide, all of which contribute to air pollution as well as water pollution when manure reaches water sources through run-off or seepage (Loyon 2018). Large-scale burning of agricultural waste (for instance, crops) also has negative health impacts through toxic smog. Indeed, the World Health Organization (2020) determines that agricultural waste burning constitutes one of the largest sources of ambient air pollution. Air pollution itself is responsible for nearly seven million human deaths annually, almost one million of which are children.

Humans also produce organic waste, when restaurants, fast-food chains, and households discard vegetables, meat, fruits, grains, and other food. According to the United Nations (2023), about thirteen percent of food is lost between harvest and retail, with another seventeen percent of food wasted in households, retail (grocery stores) and the food service industry (restaurants etc.). As the report points out, not only does all of this food loss – about thirty percent of all food produced globally – impact food insecure people, but it also contributes to GHG emissions. It also means – critically – that all of the energy and other resources devoted to *producing* the food that is wasted is also wasted. Moreover, organic waste placed in landfills creates human health and environmental problems when aerobic and anaerobic bacteria consume the organic matter and produce methane, one of the major greenhouse gases.

EXTRACTIVE INDUSTRY WASTE (FOSSIL FUELS AND MINERALS)

The further removed we are from the sources of waste, the less familiar we tend to be with the myriad types of waste that humans create. In other words, we tend to think of our global waste crisis in terms of MSW because this is the waste that we see and touch; it is the waste we participate in managing. In reality, extractive industry waste constitutes the largest waste sector (Hird 2021, 2022). Mining waste is produced when companies extract and process mineral resources (for electronics, for instance). Mining waste includes topsoil, waste rock, and mine tailings (European Commission n.d.). Some of this waste may be inert (i.e. it has no negative effect on human health or the environment) but may also contain very large quantities of dangerous substances, like heavy metals. Extracting and processing mineral resources involves the use of chemicals, which add to the often complex mix of chemical waste. Moreover, mine tailings – all of the materials left over from the process of extracting the target mineral from the rocks – are often stored in heaps exposed to the environment and/or tailings ponds, which may collapse or otherwise leak into the environment, risking human and environmental health. There are a disturbing number of examples of mine tailings disasters. For instance, on January 25th, 2019, the waste tailings pond of Brazil's Córrego do Feijão iron ore mine burst, causing a catastrophic mudslide that flowed uncontrollably for several hundred kilometers. Twelve million cubic meters of tailings made its way through five different Brazilian states, on its way to the South Atlantic Ocean, as well as two enormous hydroelectric dams. The death toll remains undetermined; authorities have confirmed 209 dead, with at least another 102 people still missing (Wilkes and Hird 2019).

China and India rank as the second and third largest consumers of oil, after the United States (International Energy Agency 2018). Extracting oil and gas from the earth also produces very significant amounts of waste, which may also be highly polluting. The Alberta oil sands alone has created over 119 km^2 of mine tailings, which leak over eleven million liters per day (Canadian Parks and Wilderness Society, Northern Alberta Chapter 2024). Well drilling produces drilling muds, cuttings, and water containing chemicals. Hydraulic fracturing (also known as 'fracking') stands to significantly increase the already 57,000 million tons of material we annually shift from beneath the earth to its surface (Douglas and Lawson 2000). Fracking is a process whereby engineered fluids containing chemicals are pumped underground at very high pressure in order to create and hold open fractures in geologic formations. The fracking fluids flow back to the surface, and must be contained in pits, tanks, or ponds. The United States Environmental Protection Agency (EPA) has thus far approved over twenty-eight different biocides for fracking applications (Kahrilas et al. 2015; Sager 2014). Biocides are chemicals that kill or otherwise render inactive microorganisms (Government of Canada 2024). Biocides may seep between rock beds, through surface soil, and into the water table. The exact chemicals that hydraulic fracturing companies use is proprietary, and not available to the public.

There is also the significant global problem of orphaned or abandoned oil and gas wells. Globally, there are an estimated twenty-nine million abandoned wells (Groom 2020). These wells may be situated in remote regions, but a significant number are located on farmland, increasing the risk to human health through land and water contamination. When fossil fuel companies no longer take responsibility for the oil and gas wells that they create, they are called orphaned wells. Sometimes wells are abandoned because they no longer create profit and/or companies declare bankruptcy and/or neglect to close or deal with (called aftercare) the wells. In the absence of strong regulations, companies save millions of dollars in cleanup and aftercare costs by abandoning the mines. As such, the funds required to close and cleanup abandoned wells becomes the responsibility of governments (i.e. taxpayer money). The transition away from fossil fuels will likely lead to a greater number of orphaned wells and their human health and environmental problems as well as tax increases. Abandoned wells are significant contributors to GHG emissions, through methane and other gases, and therefore climate change. In the United States alone, more than 3.2 million abandoned wells emitted 282 kilotons of methane in 2018: equivalent to consuming sixteen million barrels of crude oil in a typical day in the US.

PLASTICS WASTE

Plastics waste cuts across various other types of waste: a significant proportion of MSW is composed of various plastics; and plastics are also found in MLW, ICI, electronics, and even organics waste. Indeed, plastics pollution has emerged as a central global public concern.

Since the invention of the first plastics polymers over a century ago (Andrady and Neal 2009), a plethora of plastics polymers have been developed and now include: polyethylene terephthalate; high-density polyethylene; low-density polyethylene; polyvinyl chloride; polypropylene; and polystyrene. For this reason, we refer to plastics rather than plastic. Plastics producers such as PlasticsEurope promote plastics proliferation based on their many useful properties, low cost, and seemingly endless versatility for a wide variety of applications. Thus, plastics have become ubiquitous in the energy, health care, building and construction, agriculture, and transportation sectors, as well as in sports and electronics. According to *Plastics – The Facts* (PlasticsEurope 2022), we are producing more plastics than ever: from fifteen million metric tons (Mt) in 1964 increasing to over 390 Mt in 2021, and this production is expected to double within twenty years. Packaging alone accounts for almost forty percent of total plastics production.

Plastics pollution has emerged as a serious environmental concern, and is directly connected to climate change because plastics are derived from fossil fuels (EIA 2022). Plastics pollution statistics are alarming: of the more than 390 Mt of plastics produced in 2021, less than ten percent of that is recycled (PlasticsEurope 2022). Plastics take between 5 and 1,000 years to degrade into microplastics, which effectively last forever (O'Neill 2019). According to the World Economic Forum, a truck load of plastics are dumped in the world's oceans every minute (Pennington 2016) and in 2017, the United Nations declared ocean plastics to be a 'planetary crisis' with about 100,000 marine mammals dying annually due to plastics ingestion or entanglement. Unsustainable plastics production has resulted in a fifteen percent increase in GHG emissions from 2012 to 2018 (CIEL 2019). By 2050, global GHG emissions from the plastics production and consumption life cycle will account for ten to thirteen percent of our planet's remaining emissions 'budget' (CIEL 2019, 2020). Overwhelmingly (over ninety percent) of plastics are derived directly from fossil fuels: plastics and fossil fuel production increase in tandem (PlasticsEurope 2022).

Canadians produce more plastics waste per capita than anywhere else in the world: in 2019, Canada produced approximately 4.6 million Mt of plastics, while generating 2.8 million Mt of plastics waste (Deloitte 2019). Only nine percent of this country's plastics are properly recycled (Environment and Climate Change Canada 2020: 20). Even when plastics make it to recycling bins, much of it ends up in landfills while the remaining twelve percent is incinerated as waste-to-energy, producing toxic emissions into nearby communities and exacerbating GHG emissions to the atmosphere. Further, current waste sorting and processing facilities are ill-equipped to manage the flow of plastics produced from long-established plastics manufacturers (Divert NS 2021).

Packaging is the single largest source of plastics waste in the environment, and packaging is most often designed to be single-use (PlasticsEurope 2022). Our participation in plastics waste reduction is largely restricted to our role as consumers and through individual behavior change: to self-evaluating 'green' product claims, recycling, and curbside disposal (Hird 2021, 2022; Hird et al. 2014). Chapter 2 examines citizens' limited role in plastics waste management.

RADIOACTIVE WASTE

Radioactive waste is defined as any material that is radioactive or has been contaminated by radioactivity. Radioactive material is used in medicine (X-ray technology, for instance), nuclear research, nuclear power, and nuclear weapons. Radioactive waste is generally defined by the half-life of the radioactive material, which means how long (minutes, hours, years, decades, centuries, or millennia) it takes for half of the radionuclides of that material to decay. Uranium-238, for instance, has a half-life of 4.5 billion years. Strontium-90 has a half-life of twenty-nine years (Caltech n.d.).

Low-level radioactive waste (called low-level waste) includes materials that have small amounts of short-lived radioactivity, such as tools, clothing, or medical swabs. Intermediate-level radioactive waste is produced by nuclear power plants and other industries, and requires specific containment so that it does not come into contact with humans and the environment. High-level radioactive waste is highly radioactive and requires both constant cooling and permanent containment.

Nuclear power is staging a remarkable return to the energy market due to the global climate crisis. Nuclear energy is being advanced as a way of meeting increasing global energy 'needs' while decreasing fossil fuel reliance. The G7 has committed to the goal of a predominantly decarbonized electricity supply by 2035. Small modular reactor (SMR) technologies are being promoted as low-cost, safe, easily fabricated and built, flexible, and a scalable alternative to conventional nuclear reactors, producing up to 300 MWe (megawatt electrical) compared to the 700+ MWe of large, conventional reactors (Reitsma and Subki 2020). SMR technologies promise a broad sweep of applications, from replacement of large-scale diesel generators for extractive and heavy industries to on-grid power to urban and remote communities (Waters and Didsbury 2012; Canadian Small Modular Reactor Roadmap Steering Committee 2018; Fraser et al. 2022). Today, there are some eighty designs under development in nineteen countries, with two operational units – in Russia and China (Nuclear Engineering International 2023; Ingersoll 2009; Steinbruner 2014; National Energy Board 2018; Zohuri 2019; Reitsma and Subki 2020; Carelli and Ingersoll 2021). France, the Czech Republic, and Finland have announced plans to collaborate on SMR licensing, and France announced it will build six new SMRs and will consider building eight more, offering $1.1 billion for the French-owned NUWARD SMR project (Patel 2022).

The growing interest in expanded nuclear technology is echoed by the principle intergovernmental and international groups promoting climate strategies, including: the Intergovernmental Panel on Climate Change (IPPC); United Nations Environment Program (UNEP); World Meteorological Organization (WMO); the Green Climate Fund (GCF); and Climate Investment Funds (CIF). And recent developments such as the European Union (EU) proposal to classify nuclear energy as 'green' according to its action plan on sustainable growth (European Commission 2022; World Nuclear News 2022) signal the regulatory and policy uptake of what proponents are calling 'new nuclear'. The United Kingdom, the United States, France, and Canada, are keenly interested in SMRs, and Argentina, China, and Russia are currently in various stages of new SMR construction.

In the wake of the Fukushima nuclear power plant disaster in 2011, Japan declared a move away from nuclear power (but has since re-opened Fukushima and other nuclear power plants). Germany committed to phase out its remaining three reactors by the end of 2022 – and successfully completed this pledge in April 2023 (Sorge, Delfs, and Nienaber 2022; Schultheis 2023). Global uranium markets entered a decade-long decline, and many new reactor projects were postponed or cancelled. As of 2022, there are approximately 440 functioning nuclear power reactors in thirty-two countries with another fifty-five reactors under construction in nineteen countries, notably China, India, Russia, and the United Arab Emirates (World Nuclear Association 2022a), producing ten percent of global electrical energy (Ramama 2022; World Nuclear Association 2022b). Roughly two-thirds of today's operating reactors are scheduled for retirement by 2030 (IAEA 2022). And throughout this period, the many problems associated with the long-term disposition of high-level nuclear wastes have remained unsolved. To date, several nations have geological repositories in development and testing (e.g. Finland, Sweden, France, Switzerland), and other nations are in the process of locating potential sites (e.g. Canada, United States, China, Holland) (Braden and Macfarlane 2023; IAEA 2022).

Nuclear detonations have produced environmental contamination and negative human-health effects (including death, various cancers, and so on). Since the first nuclear detonation on 15 July 1945, there have been some 2,056 nuclear detonations across the globe, most of them exploded on colonized Indigenous lands far away from the capitals of the colonizing nation (Arms Control Association 2020). Nuclear testing has had carcinogenic and other negative health effects on local residents and caused considerable damage to the environment.

MILITARY/MUNITIONS WASTE

The category of military waste is arguably the most difficult to detail because of the lack of freely accessible data on this type of waste. Military waste includes the radioactive waste created by the nuclear detonations carried out by the United States, France, and other countries (see Chapter 5). Military waste also includes all other types of ammunition, such as missiles, grenades, land mines, bullets, warheads, armor-piercing shells, tracer ammunition, artillery, tanks, and so on. And it also includes all of the infrastructure, equipment, electronics, fleets, and machinery that military aircraft, land craft, and seacraft use in their conflict as well as conflict-preparedness operations (surveillance etc.). Militaries around the world participate in constant training exercises in order to maintain conflict readiness (Reno 2019).

Countries do not typically report the waste that their militaries produce, and this data is only partially available in countries that have Freedom of Information legislation (and even then, much of the information is inaccessible). The reasoning is that to do so would reveal types of weapons used and other information that may be valuable to enemy forces. Thus, information about the waste and pollution that militaries produce is largely gleaned only some years or decades after a military operation has taken place, a military base has closed, and so on. This means that much of the data is historical.

The United States has the largest military force in the world. US military bases in their own and other countries around the world are sites of significant pollution that negatively affects the health of nearby communities, as well as the environment (see Hird 2021 for Canadian examples). There are numerous active and closed military bases in the United States alone (Roels, Smith, and St. Clair 2017). Nine hundred of these sites are so contaminated that the US Environmental Protection Agency has designated them as Superfund sites, which means that they are so hazardous that the federal government assumes responsibility for remediating the sites, using taxpayer money. One of the major concerns are polyfuorinated alkyl substances (PFAS), also known as 'forever chemicals', because they do not degrade or break down easily, and thus remain in the environment, essentially, forever, without proper remediation. In 2019, the US Air Force contractor Hughes Aircraft (later Raytheon Missile Systems Co.) was found to have been releasing TCF (trichloroethylene) on the ground surrounding the Tucson International Airport for twenty-nine years, causing cancer and other illnesses amongst thousands of local residents (Rosane 2018; see also The Cost of War Project: https://watson.brown.edu/costsofwar). This is one of several concerning examples.

The extensive and intensive environmental destruction caused by war has been termed ecocide (Ziegler 2011; Giovanni 2022). The US military, as well as other militaries, have used scorched earth strategies to destroy anything that enemies might find useful, including water, land, food, infrastructure, and so on. While the 1977 Geneva Convention prohibits this policy and practice, it is still commonly used, with devastating environmental consequences as it creates wastelands, such as in the Native American wars in the (now) United States, World Wars I and II, and the Vietnam War. The US and other militaries have also used various chemicals, such as Agent Orange in Vietnam, that have negatively impacted the environment. In Vietnam alone, the US military deployed over '20 million gallons of herbicides… to defoliate forests, clear growth along the borders of military sites and eliminate enemy crops' (King 2006). These herbicides devastated wildlife as well: of the 145–170 bird species that existed before the war, only twenty-four species remain; and of the thirty to fifty-five kinds of mammals, only five species remain (DeWeerdt 2008).

Wars in Rwanda and many other countries have created vast stockpiles of waste and pollution. Some of this waste has been both legally and illegally disposed of over land and in oceans. For instance, in the aftermath of World War II, in 1946 the Canadian military dumped 10,982 forty-five-gallon barrels of mustard gas into the ocean, 180 miles south-east of Halifax, at a depth of some 1,350 fathoms. The mustard gas was manufactured by Stormont Chemicals, in a secret warfare factory in Cornwall, Ontario (Kehoe 2002). The United Kingdom, the United States, and other countries dumped mustard gas as well as munitions and other contaminated materials off their coasts as well. Indeed, the Canadian Air Force's 'Military Top Priority One Classification' includes twenty-eight sites 'known to have reactive and viable chemical and biological weapons or nuclear weapons'. The location of ten of these sites appears on local fishing charts; the location – and what each site contains – of the remaining eighteen remain unknown, even to the military. And even when the military does know the location, 'security and safety considerations could preclude the release of the exact locations due to the materials that may be present at these sites', states John McCallum, Minister of National Defence. Greatly understating the problem, Wendell King, former Colonel in the US Army, wrote, 'Our legacy of environmental protection and military installations is not the most exemplary part of our military history' (King 2001: x).

Military waste also encompasses all of the waste produced through the use of fossil-fuel-based vehicles and machinery. The United States' decision to withdraw from the 1997 Kyoto Protocol means that it is exempt from reporting even their military carbon emissions (Lewis 2021). Researchers have found that 'the US military is one of the largest climate polluters in history, consuming more liquid fuels and emitting more CO_2e (carbon dioxide equivalent) than most countries' (Belcher et al. 2019: 65). Indeed, the US military alone emits more carbon than whole countries such as Morocco, Peru, Sweden, Hungary, Finland, New Zealand, Norway, and Switzerland (Lewis 2021). In 2017, the US military purchased some 269,230 barrels of oil *per day*, and emitted more than 25,000 kt CO_2e. In the same year, the US Air Force purchased $4.9 billion, the Navy $2.8 billion, the Army $947 million and the Marines $36 million worth of fuel.

REVIEW

There are numerous kinds of waste that countries and organizations group into different categories according to shifting priorities. People are most familiar with the waste that they regularly come into contact with: municipal solid and liquid waste. These categories of waste, while increasing in both volume and diversity of materials, are miniscule in comparison to a much greater labyrinth of waste produced in the extraction, manufacturing and retail phases of any given product's life-cycle. It is difficult to get information about the material cost of the complete life-cycle of products: manufacturers are not required to provide this information, and estimating it is difficult (but not impossible). What is clear is that volumes of all kinds of waste are increasing, despite government and industry efforts to reduce waste (mainly through recycling programs). Chapter 2 examines how and why waste has become a global human health and environmental concern by detailing affluent countries' significant increase in energy and product production and consumption, both of which have resulted in greater volumes of waste.

SUGGESTED READING

Franklin-Wallis, O. (2023). *Wasteland: The Secret World of Waste and the Urgent Search for a Cleaner Future*. Hachette Books.

O'Neill, K. (2019). *Waste*. Polity Press.

Rogers, H. (2005). *Gone Tomorrow: The Hidden Life of Garbage*. The New Press.

Royte, E. (2005). *Garbage Land: On the Secret Trail of Trash*. Little, Brown and Company.

REFERENCES

ANDRA. (n.d.). Waste Classification. https://international.andra.fr/radioactive-waste-france/waste-classification. Accessed 4 October 2023.

Andrady, A.L. and Neal, M.A. (2009). Applications and Societal Benefits of Plastics, *Philosophical Transactions of the Royal Society B: Biological Sciences*, *364*:1977–1984.

Arms Control Association. (2020). The Nuclear Testing Tally. www.armscontrol.org/factsheets/nucleartesttally. Accessed 31 December 2020.

Belcher, O., Bigger, P., Neimark, B., and Kennelly, C. (2019). Hidden Carbon Costs of the 'Everywhere War': Logistics, Geopolitical Ecology, and the

Carbon Boot-Print of the US Military, *Transactions of the Institute of British Geographers*, *45* (*1*): 65–80.

Berendes, D.M., Yang, P.J., Lai, A., Hu, D., and Brown, J. (2018) Estimation of Global Recovery of Human and Animal Faecal Biomass, *Nature Sustainability*, *1*: 679–685.

Blaustein-Rejto, D. (2023). Livestock Don't Contribute 14.5% of Global Greenhouse Gas Emissions. The Breakthrough Institute, 20 March. https://thebreakthrough.org/issues/food-agriculture-environment/livestock-dont-contribute-14-5-of-global-greenhouse-gas-emissions. Accessed 25 June 2024.

Braden, Z. and Macfarlane, A.M. (2023). The Final Countdown to Site Selection for Canada's Nuclear Waste Geologic Repository, *Bulletin of the Atomic Scientists*, *79* (*1*): 22–27.

Bureau of Transportation Statistics. (2020). Municipal Solid Waste and Construction and Demolition Debris. www.bts.gov/archive/subject_areas/freight_transportation/faf/faf4/debris. Accessed 4 October 2023.

Caltech. (n.d.). *Radioactive Nuclei with a Half-Life Greater Than 1000 Years*. https://sites.astro.caltech.edu/~dperley/public/isotopetable.html. Accessed 12 August 2024.

Canadian Parks and Wilderness Society, Northern Alberta Chapter. (2024). *Oil Sands Tailings*. https://cpawsnab.org/our-work/oil-sands-tailings/. Accessed 17 June 2024.

Canadian Small Modular Reactor Roadmap Steering Committee. (2018). *A Call to Action: A Canadian Roadmap for Small Modular Reactors*. NRCAN.

Carelli, M.D. and Ingersoll, D.T. (2021). *Handbook of Small Modular Nuclear Reactors* (2nd edn). Woodhead Publishing.

Carrington, D. (2020) $10bn of Precious Metals Dumped Each Year in Electronic Waste, Says UN, *The Guardian*, 2 July. www.theguardian.com/environment/2020/jul/02/10bn-precious-metals-dumped-each-year-electronic-waste-un-toxic-e-waste-polluting. Accessed 7 October 2023.

Center for Disease Control. (2022). *Water Treatment*. www.cdc.gov/healthy-water/drinking/public/water_treatment.html. Accessed 17 June 2024.

Chong, W.K. and Hermreck, C. (2010) Understanding Transportation Energy and Technical Metabolism of Construction Waste Recycling, *Resources, Conservation and Recycling*, *54* (*9*): 579–590.

CIEL (Centre for International Environmental Law). (2019). *Plastic and Climate: The Hidden Costs of a Plastic Planet*. Accessed 7 December 2020. www.ciel.org/wp-content/uploads/2019/05/Plastic-and-Climate-FINAL-2019.pdf.

CIEL (Centre for International Environmental Law). (2020). Convention on Plastic Pollution: Toward a New Global Agreement to Address Plastic Pollution. www.ciel.org/reports/convention-on-plastic-pollution-toward-a-new-global-agreement-to-address-plastic-pollution/. Accessed 14 May 2021.

Circle Economy. (2023). *The Circularity Gap Report 2023*. www.circle-economy.com/news/cutting-material-consumption-by-one-third-is-key-to-tackling-climate-change-study. Accessed 25 June 2024.

Connecticut Department of Energy and Environmental Protection. (2020). *Climate Change and Waste*. https://portal.ct.gov/DEEP/Reduce-Reuse-Recycle/Climate-Change/Climate-Change-and-Waste#:~:text=Our%20Wasteful%20Impact%20on%20Climate%20Change&text=Solid%20waste%20contributes%20directly%20to,our%20solid%20waste%20combustion%20facilities. Accessed 26 September 2023.

Deloitte. (2019). *Economic Study of the Canadian Plastic Industry, Market and Waste Task 5 – Summary Report to Environmental and Climate Change Canada*. Deloitte LLC.

DeWeerdt, S. (2008). War and the Environment, *World Wide Watch, 21 (1)*. www.envirosagainstwar.org/2015/04/09/war-and-the-environment-3/. Accessed 22 October 2024.

Divert NS. (2021). *Sorting Guides*. https://divertns.ca/recycling/sorting-guide. Accessed 19 October 2024.

Douglas, I. and Lawson, N. (2000). The Human Dimensions of Geomorphological Work in Britain, *Journal of Industrial Ecology, 4 (2)*: 9–33.

Environment and Climate Change Canada. (2020). *A Proposed Integrated Management Approach to Plastic Products to Prevent Waste and Pollution*. Discussion paper. Government of Canada. www.canada.ca/content/dam/eccc/documents/pdf/cepa/proposed-approach-plastic-management-eng.pdf. Accessed 19 October 2024.

Environmental Investigation Agency (EIA). (2022). *Connecting the Dots: Plastic Pollution and the Planetary Emergency*. https://eia-international.org/wp-content/uploads/2022-EIA-Report-Connecting-the-Dots-SPREADS.pdf. Accessed 12 November 2022.

European Commission. (2022) EU Taxonomy for Sustainable Activities. https://ec.europa.eu/info/business-economy-euro/banking-and-finance/sustainable-finance/eutaxonomy-sustainable-activities_en. Accessed 17 March 2022.

Fraser, P., Wanner, B., Everhart, K., and Herzog, A. (2022). *Nuclear Power and Secure Energy Transitions: From Today's Challenges to Tomorrow's Clean Energy Systems*. International Energy Agency. https://iea.blob.core.windows.net/assets/0498c8b8-e17f-4346-9bdedad2ad4458c4/NuclearPowerandSecure-EnergyTransitions.pdf. Accessed 19 October 2022.

Garcier, R. (2014). Disperse, Confine, or Recycle? A Geo-logical Approach to the Management and Spatial Circulations of Low-level Radioactive Waste in France, *L'Espace Géographique, 43 (3)*: 265–283.

Giovanni, C. (2022). Ecocide: From the Vietnam War to International Criminal Jurisdiction? Procedural Issues in Between Environmental Science, Climate

Change, and Law. *Cork Online Law Review.* https://papers.ssrn.com/sol3/papers.cfm?abstract_id=4072727. Accessed 13 August 2024.

Government of Canada. (2024). *Biocides.* www.canada.ca/en/health-canada/services/drugs-health-products/biocides.html#. Accessed 12 August 2024.

Groom, N. (2020). Special Report: Millions of Abandoned Oil Wells are Leaking Methane, a Climate Menace. *Reuters, 16 June.* www.reuters.com/article/us-usa-drilling-abandoned-specialreport-idUSKBN23N1NL. Accessed 4 October 2023.

Hird, M.J. (2021). *Canada's Waste Flows.* McGill-Queen's University Press.

Hird, M.J. (2022). *A Public Sociology of Waste.* Bristol University Press.

Hird, M.J., Lougheed, S., Rowe, K. and Kuyvenhoven, C. (2014). Making Waste Management Public (or Falling Back to Sleep), *Social Studies of Science, 44* (*3*): 441–465.

Hoesly, R.M. et al. (2018). Historical (1750–2014) Anthropogenic Emissions of Reactive Gases and Aerosols from the Community Emissions Data System (CEDS). *Geoscientific Model Development, 11*, 369–408.

Howell, B. (2022) The 7 Biggest Polluters by Industry in 2022, as Ranked in New Research, *Environmental Protection,* 17 October. https://eponline.com/articles/2022/10/14/the-7-biggest-polluters.aspx. Accessed 4 October 2023.

Ingersoll, D.T. (2009). Deliberately Small Reactors and the Second Nuclear Era. *Progress in Nuclear Energy, 51* (*4–5*), 589–603.

Intergovernmental Panel on Climate Change (IPPC). (2018). *Summary for Policymakers of IPCC Special Report on Global Warming of 1.5°C.* www.ipcc.ch/sr15/chapter/spm/. Accessed 8 October 2023.

International Atomic Energy Agency (IAEA). (2022). *Status and Trends in Spent Fuel and Radioactive Waste Management.* IAEA.

International Energy Agency. (2018). *World Energy Outlook 2018.* IEA.

Kahrilas, G.A., Blotevogel, J., Stewart, P.S., and Borch, T. (2015). Biocides in Hydraulic Fracturing Fluids: A Critical Review of Their Usage, Mobility, Degradation, and Toxicity, *Environmental Science and Technology, 49* (*1*): 16–32.

Kaza, S., Yao, L., Bhada-Tata, P., and Van Woerden, F. (2018). *What a Waste 2.0: A Global Snapshot of Solid Waste Management to 2050.* Urban Development Series. World Bank. https://doi.org/10.1596/978-1-4648-1329-0.

Kehoe, M. (2002). Military Dumpsites off Canada's Atlantic Coast. Office of the Auditor General of Canada, Petition: No. 50A. www.oag-bvg.gc.ca/internet/English/pet_050A_e_28755.html. Accessed 19 October 2024.

King, J. (2006). Vietnamese Still Paying a High Price for Chemical Warfare, *The Independent,* 8 July. www.independent.co.uk/climate-change/news/vietnamese-wildlife-still-paying-a-high-price-for-chemical-warfare-5329662.html. Accessed 13 August 2024.

King, W.C. (2001). Foreword. In Ehlen, J. and Harmon, R.S. (eds.), *The Environmental Legacy of Military Operations,* (Reviews in Engineering Geology, Vol. *XIV*). The Geological Society of America.

Lewis, J. (2021). US Military Pollution: The World's Biggest Climate Change Enabler? *Earth.org*, 12 November. https://earth.org/us-military-pollution/. Accessed 13 August 2024.

Loyon, L. (2018). Overview of Animal Manure Manufacture for Beef, Pig and Poultry Farms in France, *Frontiers in Sustainable Food Systems*, *2*. doi:10.3389/fsufs.2018.00036.

Martinez, C. (2020) Retail Waste Adds Up to Big Losses for Merchants. *Signifyd*. www.signifyd.com/blog/retail-waste-losses/. Accessed 4 October 2023.

McKendrik, J. (2022). They're No Longer Cars, They're Computers on Wheels, *RT Insights*. www.rtinsights.com/theyre-no-longer-cars-theyre-computers-on-wheels/#:~:text=Edge%20and%205G%20computing%20are,truly%20are%20computers%20on%20wheels. Accessed 7 October 2023.

Mester, T., Szabó, G., Sajtos, Z., Baranyai, E., Szabó, G., and Balla, D. (2022). Environmental Hazards of an Unrecultivated Liquid Waste Disposal Site on Soil and Groundwater, *Water*, *14* (*2*): 226.

National Energy Board. (2018). *Nuclear Energy in Canada: Energy Market Assessment*. www.cer-rec.gc.ca/en/data-analysis/energy-commodities/electricity/report/archive/2018-nuclear-energy/2018nclrnrg-eng.pdf. Accessed 25 June 2024.

National Geographic. (2024). *Great Pacific Garbage Patch*. https://education.nationalgeographic.org/resource/great-pacific-garbage-patch/. Accessed 17 June 2024.

Novotny, T.E. and Zhao, F. (1999). Consumption and Production Waste: Another Externality of Tobacco Use. *Tobacco Control*, *8*: 75–80.

Nuclear Engineering International. (2023). Replacing Russia. *NEI Magazine*, 20 September. www.neimagazine.com/advanced-reactorsfusion/replacing-russia-11162298/. Accessed 25 June 2024.

OECD. (2015) *OECD-FAO Agricultural Outlook 2015*. www.oecd-ilibrary.org/agriculture-and-food/oecd-fao-agricultural-outlook-2015/per-capita-meat-consumed-in-the-world_agr_outlook-2015-graph94-en. Accessed 8 October 2023.

O'Neill, K. (2019). *Waste*. Polity Press.

Optoro. (2022). *2022 Impact Report: Making Returns Better for Customers, Retailers, and the Planet*. https://4771362.fs1.hubspotusercontent-na1.net/hubfs/4771362/2022%20Impact%20Report/Optoro_2022%20Impact%20Report.pdf?utm_campaign=Impact%20Report%202021&utm_medium=email&_hsmi=245224656&_hsenc=p2ANqtz-8jFcrcynZu7--xYwyD6fvozxCs8bd4V4zTPQrlXM--XjHB8mZot4DHoxLxaNPf1FiEL-W52AFPDsGPzSD3Z1JWIT80BefkIkE8ZgGjsY7mrpP6hpzw&utm_content=245223033&utm_source=hs_email. Accessed 4 October 2023.

Our World in Data. (n.d. a). *Daily Meat Consumption Per Person 2020*. https://ourworldindata.org/grapher/daily-meat-consumption-per-person. Accessed 8 October 2023.

Our World in Data. (n.d. b). *Feed Required to Produce 1 Kilogram of Meat*. https://ourworldindata.org/grapher/feed-required-to-produce-one-kilogram-of-meat-or-dairy-product. Accessed 8 October 2023.

Patel, S. (2022). France's NUWARD SMR Will Be Test Case for European Early Joint Nuclear Regulatory Review, *Power Magazine*, 2 June. www.powermag.com/frances-nuward-smr-will-be-test-case-for-european-early-joint-nuclear-regulatory-review/. Accessed 25 June 2024.

Pennington, J. (2016). *Every Minute, One Garbage Truck of Plastic Is Dumped into Our Oceans. This Has to Stop.* www.weforum.org/agenda/2016/10/every-minute-one-garbage-truck-of-plastic-is-dumped-into-our-oceans/. Accessed 18 December 2020.

PlasticsEurope. (2022). *Plastics – The Facts 2022*. PlasticsEurope AISBL.

Ramama, M.V. (2022). The Hollow Promise of Small Modular Nuclear Reactors. *Counterpunch, August 3*.

Reitsma, F. and Subki, H. (2020) *Advances in Small Modular Reactor Technology Developments. IAEA-NPTD Webinar on Advances in Small Modular Reactor (SMR) Design and Technology Developments: A Booklet Supplement to the IAEA Advanced Reactors Information System (ARIS)*. International Atomic Energy Agency

Reno, J. (2019). *Military Waste: The Unexpected Consequences of Permanent War Readiness*. University of California Press.

Roels, C., Smith, B., and St. Clair, A. (2017). Military Bases' Contamination Will Affect Water for Generations, *The Center for Public Integrity*, 18 August. https://publicintegrity.org/environment/military-bases-contamination-will-affect-water-for-generations/. Accessed 13 August 2024.

Rosane, O. (2018). Pollution from Air Force Keeps Causing Cancer in Tucson, Residents Say, *EcoWatch*, 27 May. www.ecowatch.com/tucson-pollution-air-force-cancer-2553583119.html. Accessed 13 August 2024.

Sager, J. (2014). Fracking Floods the Earth with Biocides, *The Progressive Cynic*, May 19. http://theprogressivecynic.com/2014/05/19/fracking-floods-the-earth-with-biocides/. Accessed 19 October 2024.

Schecht, S., Real, E., Létinois, L., Colette, A., Holland, M., Spadaro, J.V., Opie, L., Brook, R., Garland, L., Gibbs, M., Calero, J., Zeiger, B., Rouïl, L., Brignon, J.M., and German, R. (2020). *ETC/ATNI Report 04/2020: Costs of Air Pollution from European Industrial Facilities 2008–2017*. European Topic Centre on Air Pollution, Transport, Noise and Industrial Pollution.

Schultheis, E. (2023). Germany Turns Out the Lights on Nuclear Power – at Last, *Foreign Policy*, 15 April. https://foreignpolicy.com/2023/04/15/germany-nuclear-power-shutdown-energy-policy. Accessed 25 June 2024.

Smith, R.J. (1982) The Risks of Living Near Love Canal, *Science, 217*: 808–809, 811.

Sorge, P., Delfs, A., and Nienaber, M. (2022). Scholz Extends Germany's Last Nuclear Plants to Quell Feud, *Bloomberg Press*, 16 October. www.bloomberg.

com/news/articles/2022-10-17/germany-to-extend-lifetime-of-all-three-nuclear-power-plants?embedded-checkout=true. Accessed 25 June 2024.

Statista. (n.d.). *Global E-Waste: Statistics and Facts*. www.statista.com/topics/3409/electronic-waste-worldwide/#topicOverview. Accessed 7 October 2023.

Statista. (2023). *Market Size of Waste Management Worldwide in 2022, with Forecasts Until 2030*. www.statista.com/statistics/246178/projected-global-waste-management-market-size/. Accessed 12 July 2023.

Steinbruner, J. (2014). Anticipating Climate Mitigation: The Role of Small Modular Nuclear Reactors (SMRs). Working Paper. Center for International and Security Studies, U. Maryland. www.jstor.org/stable/resrep04991. Accessed 27 May 2022.

Straková, J., Brosché, S. and Brabcová, K. (2023). *Toxics in our Clothing: Forever Chemicals in Jackets and Clothing from 13 Countries*. IPEN and Arnika.

Straková, J., Schneider, J. and Cingotti, N. (2021). *Throwaway Packing, Forever Chemicals: European Wide Survey of PFAS in Disposable Food Packaging and Tableware*. Anika Organization. https://arnika.org/en/publications/throwaway-packaging-forever-chemicals-european-wide-survey-of-pfas-in-disposable-food-packaging-and-tableware. Accessed 17 June 2024.

United Nations. (2012). *The Global Garbage Crisis: No Time to Waste*. www.unep.org/news-and-stories/press-release/global-garbage-crisis-no-time-waste. Accessed 12 October 2022.

United Nations. (2023). *Reducing Food Loss and Waste: Taking Action to Transform Food Systems*. www.un.org/en/observances/end-food-waste-day. Accessed 8 October 2023.

United Nations Environment Programme (UNEP). (2024). *Beyond an Age of Waste: Turning Rubbish into Resource*. https://wedocs.unep.org/bitstream/handle/20.500.11822/44939/global_waste_management_outlook_2024.pdf?sequence=3. Accessed 12 August 2024.

United States Environmental Protection Agency. (2022). *Containers and Packaging: Product-Specific Data*. www.epa.gov/facts-and-figures-about-materials-waste-and-recycling/containers-and-packaging-product-specific. Accessed 4 October 2023.

Ward, C.P., Armstrong, C.J., Walsh, A.N., Jackson, J.H., and Reddy, C.M. (2019) Sunlight Converts Polystyrene to Carbon Dioxide and Dissolved Organic Carbon, *Environmental Science Technology Letters*, 6 (11): 669–674.

Waters, C. and Didsbury, R. (2012). Small Modular Reactors – A Solution for Canada's North? *AECL Nuclear Review*, 1 (2): 3–7.

Wilkes, J. and Hird, M.J. (2019) Colonial Ideologies of Waste: Implications for Land and Life, *EuropeNowJournal*, 27 (May). www.europenowjournal.org/2019/05/06/confronting-waste/. Accessed

World Health Organization. (2018). *Health-Care Waste*. www.who.int/news-room/fact-sheets/detail/health-care-waste. Accessed 4 October 2023.

World Health Organization. (2020). *Ambient (Outdoor) Air Pollution*. www. who.int/news-room/fact-sheets/detail/ambient-(outdoor)-air-quality-and-health. Accessed 8 October 2023.

World Meteorological Organization. (2015). *Global Climate in 2015–2019: Climate Change Accelerates*. https://wmo.int/news/media-centre/global-climate-2015-2019-climate-change-accelerates. Accessed 1 November 2023.

World Nuclear Association. (2022a). *Plans For New Reactors Worldwide*. https:// worldnuclear.org/information-library/current-and-future-generation/plans-for-new-reactors-worldwide.aspx. Accessed 19 October 2024.

World Nuclear Association. (2022b). *Ukraine: Russia-Ukraine War and Nuclear Energy*. https://world-nuclear.org/information-library/countryprofiles/countries-t-z/ukraine-russia-war-and-nuclear-energy.aspx. Accessed 19 October 2024.

World Nuclear News. (2022). *A Guide to the EU's 'Green' Taxonomy – and Nuclear's Place in It: Energy & Environment*. www.world-nuclearnews.org/Articles/A-guide-to-the-Eus-green-taxonomy-and-nuclears-pla. Accessed 19 October 2024.

Wynne, B. (1987). *Risk Management and Hazardous Waste: Implementation and Dialectics of Credibility*. SpringerVerlag.

Ziegler, D. (2011). *The Invention of Ecocide: Agent Orange, Vietnam, and the Scientists who Changed the Way We Think About the Environment*. University of Georgia Press.

Zohuri, B. (2019). *Small Modular Reactors as Renewable Energy Sources*. Springer International Publishing.

WASTE IN HISTORICAL CONTEXT

OVERVIEW

Chapter 2 details our current global waste crisis in historical context. Until relatively recently, individuals and individual cartage and disposal operators handled municipal solid waste, removing it from curbsides and transporting it to open dumps, and later, engineered landfills, incinerators, and energy-from-waste facilities. Now, large corporations manage waste in affluent countries, while waste is still openly dumped in poorer regions of the world. Two major historical and contemporary forces – capitalism and colonialism – have had, and continue to have, profound effects on waste production and management. Chapter 2 focuses on these forces, and the staggering increase in global extraction, production, and consumption as the foundation of modern capitalist economies.

INTRODUCTION

The history of waste reveals a slow but momentous shift from informal to formal waste management. Countries with sufficient resources (including funds and infrastructure) have introduced modern waste management systems that consist of waste sorting (primarily for disposal and recycling), transport, and disposal in engineered landfills and/or incinerators, and waste exporting to other regions or countries.

DOI: 10.4324/9781003398424-2

The waste produced by fossil fuels and mining companies has historically been largely unregulated. Currently, regulations governing producer waste are often incomplete and/or are self-reporting. Major steps need to be taken globally to much more effectively regulate and enforce regulations pertaining to resource extraction and producer industries. Moreover, a significant part of modern waste management involves exporting waste to other regions of the world, including countries that do not have the resources to refuse waste imports.

PRE-CAPITALISM: WASTE NOT, WANT NOT

Since humans have gathered in communities, we have both produced and managed waste. In early societies, a significant proportion of waste consisted of biodegradable materials such as animal bones, hides, and shells, which were sometimes dumped in waterways or on land. Other non-biodegradable materials, such as pottery, were burned or dumped together. Archaeologists have found open dumpsites – called middens – across the globe, and often situated on the outskirts of ancient communities. For instance, researchers discovered a two-meter deep midden on one of the remote Lizard Islands off the coast of Queensland, Australia (Australian Research Council 2022). The midden contains the bones of numerous animal and fish and shellfish species, and their analysis is helping scientists understand how Aboriginal peoples over 6,000 years ago interacted with their environment.

Syria created the first known wastewater management system around 6500 BCE, with sophisticated drainage, gutters, and chambers for sediment deposits. Other wastewater systems have been found in ancient Mesopotamia, Egypt, Greece, Iran, and Italy (Mays 2010). Some societies such as the Minoans (1500 BCE) had more organized practices of waste-dumping outside the Cretan capital of Knossos. The ancient Roman Empire's famous Cloaca Maxima consisted of an extensive sewer network that drained into the Tiber River, evacuating the city of its human excrement while polluting the river. Once the Romans determined that the Tiber water was causing illnesses, they developed aqueduct technology to ensure potable drinking water. Athens instituted the first known law prohibiting the practice of open dumping in 320 BCE. But well into the Middle Ages, as more and more people moved from communities that grew into towns and then cities in Europe, waste was managed in a fairly haphazard way, with people dumping waste out of their windows onto the ground below, feeding organic food scraps to animals such as pigs that lived within or near to households, and taking waste on carts to open dumpsites. Access to sewer systems, clean drinking water, and waste collection services were reserved for those who could pay.

Historical studies have produced some interesting, if rather disturbing, accounts of waste dumping in European cities. For example, Nor Loch was a human-made lake that King James III ordered to be constructed in 1460 in what was known as Old Town in Edinburgh, Scotland. During the Middle Ages, the population of Old Town grew rapidly and Nor Loch was polluted with discarded material thrown into the lake from the townsfolk, as well as sewage. It also contained human remains. Nor Loch was a popular place for suicide attempts as well as executions in the form of drowning. Successive municipal workers have discovered large chests containing human skeletons (Fife 2004). European industrialization brought even more people to cities, and as populations and crowding grew, so did the build-up of waste and its human health consequences.

SCAVENGING AND THRIFTING

People combing through trash for reusable and/or profitable material is a centuries' old practice. In England, for instance, 'dust collectors' collected coal ash and then sold it for brick-making and agricultural soil enhancement (Velis, Wilson, and Cheeseman 2009). Up until the mid-twentieth century, people in economically developed countries like France, the United Kingdom, and the United States regularly reused, mended, refurbished, and otherwise held on to their clothes, household furnishings and food leftovers for as long as possible. Factories at this time commonly reserved some jobs for sorting through ashes, bones, and other materials left over during the manufacturing process. Thrifting, reusing and refurbishing – and just plain doing without – has largely been replaced in affluent nations in what is termed our 'throwaway society' (Strasser 1999). Now these things – sewing, knitting, and combing through clothes in a used clothing store – are defined as hobbies rather than vital to household economic maintenance, although there is an increasing blurring of this boundary as the middle-class shrinks and the income divide between wealthy and poor widens.

But before the twentieth century, consuming less was normal. As well as middle-class people trading rags and other materials to peddlers – men who went door-to-door in communities collecting these materials and then selling them to buyers – in exchange for needed items, people:

> practiced habits of reuse that had prevailed in agricultural com-munities here and abroad. Women boiled food scraps into soup or fed them to domestic animals; chickens, especially, would eat almost anything and return the favor with eggs. Durable items were passed on to people of other classes or generations, or stored in attics or basements for later use. Objects of no use to adults became playthings for children. Broken or worn-out things could be brought back to their makers, fixed by somebody handy, or taken to people who specialized in repairs. And items beyond repair might be dismantled, their parts reused or sold to junk men who sold them to manufacturers. Things that could not be used in any other way were burned; especially in the homes of the poor, trash heated rooms and cooked dinners.
>
> (Strasser 1999: 12)

A thriving popular literature concerned itself with offering advice to women about how to best manage their households (including slaves and servants in the United States and other colonial nations) through thrift. Books and magazines such as Christine Herrick's *Housekeeping Made Easy* and Lydia Child's *The American Frugal Housewife* extolled the moral and practical virtues of minimizing waste as much as possible, and contained all sorts of varied advice from glue-making recipes to repairing broken ceramics and marble to chemical mixtures to remove stains, waterproof fabrics, and clean chamber pots. Serving leftovers was elevated to an art form in such books as *The Family Save-All: A System of Secondary Cookery*, with a plethora of recipes and tips for re-serving food to families. When that route was exhausted, women were advised to give food to the household livestock and to servants (who, the books and magazines warned, must be properly supervised at all times lest they steal food from their employers). Women were advised how to patch, darn, and otherwise mend clothing. When one child outgrew a shirt or pants, it should be passed down to the next child, or taken apart and fabric added to enlarge the garment for further wear. Rags could be made into rugs and quilts to keep homes warm. Even the packaging that goods such as furniture came in was reused. First of all, the packaging was made out of fabric, wood, and other much more readily reusable materials (rather than single-use plastics). And as Susan Strasser (1999) points out, people paid for packaging (directly, and not as part of the cost of the product, as is now the case). The packaging was valuable, and businesses reused this packaging. Consumers also used packaging: 'the purple wrapping paper that came with sugar… could be used for dye, the brown overalls patched with flour sacks, the rugs hooked on coffee sacks, the paint-covered pickle jar' (1999: 67). As late as 1882, manuals had to actually tell readers what a wastepaper basket was. In all of this advice, women were central to the value of stewardship; of keeping their homes, husbands, and children clean and tidy. Women were also to gather up any household surplus, from crockery to coats and socks, and donate them to poorer families.

As such, up until industrialization, economies were largely circular. Extensive and everyday thrifting, reuse, and informal collectors collecting and selling back to manufacturers who in turn used these materials in their production of new products meant that raw and manufactured materials remained for much longer periods in circulation, rather than going to waste shortly after their first consumption. This is not to say that industrialization had no negative impact on the environment: on the contrary, industrialization produced extensive land, water, and air pollution. In terms of waste, two significant developments followed. The first was the development of public health and hygiene. The second was the development of private waste management within capitalist economies.

PUBLIC HEALTH

Before contemporary waste management services developed, neither rural communities nor growing cities organized public management systems for waste collection, street cleaning, water treatment, or human waste removal and treatment (Louis 2004). Various disease outbreaks that caused widespread death stimulated calls for organized public health. For instance, the cholera outbreak in London, England in the mid-nineteenth century led John Snow to identify the public water supply system as the cause of the outbreak (Tulchinsky 2018). *The Sanitary Condition of the Labouring Population* 1842 report called for organized waste removal and management in order to improve public health, particularly for England's working class, on whose labor the wealthy classes depended for profit. Similar shifts were taking place across European and other countries at this time. Gradually, sanitary engineers began working with public health authorities to create the infrastructure needed to establish sewage treatment facilities, incinerators, and other waste management technologies. Cities bought waste collection trucks, motorized street sweepers, and constructed incinerators and landfills (Louis 2004). Home-making magazines and books began including articles by health and hygiene experts that advised middle- and upper-class wives and mothers to prioritize the health of their husbands and children by throwing away articles that might decrease their loved ones' hygiene. Single-use sanitary products for girls and women replaced reusable products. Single-use paper tissues replaced washable handkerchiefs. Disposable razors replaced reusable razors, and cleaning products packaged in layers of plastics replaced reusable glass bottles.

As the costs of this infrastructure, landfill land acquisition, and expertise increased, responsibility shifted to municipalities to organize and pay for their constituents' municipal solid and liquid waste generation. Municipal and regional administration did this by paying an increasing number of private waste management companies.

PRIVATIZING WASTE MANAGEMENT

Where before peddlers could make a reasonable living for themselves and their families through informally organized 'rags' trades – positioning themselves between households and manufacturers – increasing extraction and manufacturing were producing far more consumer goods. The sheer volume of industrial waste itself – all of the waste that manufacturing produced – shifted the peddler's focus away from collecting unwanted items from individual households and towards all the waste that industry itself was increasingly producing. Mass machines produced mass products, and individual peddlers were gradually usurped by an increasing number of private companies that were specializing in waste collection. 'For the first time in human history' Strasser notes, 'disposal became separated from production, consumption, and use' (1999: 109).

Municipalities were under pressure to formally organize waste collection and removal from city streets. Gradually, municipalities introduced laws prohibiting people from tossing their garbage in the streets, and rural communities from open dumping. At this time, homes were still relatively small compared to the much larger houses that middle-class people live in now, and this meant households produced more trash as they did not have the space for their increased consumption (Strasser 1999). A few of the small peddler traders were able to grow their businesses into larger businesses, with trucks and other infrastructure, labor, and the investments needed to secure contracts with an ever-expanding number of cities. Waste Management Inc., the world's largest waste management company, tells its history as a story of the American Dream: a tale of plucky individuals who saw an opportunity and through hard work and initiative grew a company from the ground up into the empire it is today:

> [I]n 1893… a Dutch immigrant named Harm Huizenga began collecting trash in Chicago for a small fee. With little more than a modest wagon, he built up a reliable client base over the years by effectively managing the waste of a rapidly changing society… In 1968, Harm's grandson Wayne Huizenga and two other investors, Dean Buntrock and Larry Beck, had a vision. They wanted to serve their community by properly managing the waste produced by a rapidly growing population consuming more and more products built for convenience.
>
> (Waste Management Inc. n.d.)

As Chapter 4 details, this narrative of meeting consumer's needs prevails within the waste management industry, and is a key part of wealthy nations' throwaway society.

CAPITALISM

Capitalism is broadly defined as an economic system that is based on profiting from private ownership. Capitalism includes several key features, all of which apply to waste management. Capitalist accumulation refers to the investment of money or any kind of financial assets with the goal of increasing the monetary value as financial return (Caves 2004). Property owners do this when they rent to tenants. Individuals and companies do this when they invest in the stock market. Waste management companies do this when they buy land that is then used for landfilling or an incineration facility. As the International Monetary Fund explains:

> In a capitalist economy, capital assets – such as factories, mines and railroads – can be privately owned and controlled, labor is purchased for money wages, capital gains accrue to private own-ers, and prices allocate capital and labor between competing uses.
> (Jahan and Mahmud 2015: 44)

As well as private property ownership, key features of capitalism in its ideal (rather than actual) form include: self-interest whereby all individuals and companies act in ways that benefit themselves; competition such that individuals and companies compete against each other to maximize their own interests and profits; a market mechanism through which prices are determined through maxi-mum competition between buyers and sellers; the freedom to choose such that companies can choose where it is most profitable to operate, consumers can choose to buy the products they want, and employees can choose to stay or leave their jobs; and the lim-ited role of government of intervening in the aforementioned key features of capitalism.

As Chapter 5 details, colonialism was (and remains) profoundly intertwined with capitalism in its global reach. Colonizing coun-tries, including Portugal, Spain, the Netherlands, England, Italy, and France developed as capitalist nations, in large part, through their colonial ambitions. Through violent dispossession and con-trol, colonizing nations accumulated wealth and political stability at home, while also establishing capitalist institutions in colonized nations (Austin 2014). In the nearly 500-year continuous history of colonialism, slave and indentured labor and new economic markets were central to capitalism. As Gareth Austin (2014) notes, slave and indentured labor afforded colonizers maximum profit, with laborers unable to freely leave, negotiate wages, and so on. Slave and inden-tured labor enabled colonizing countries to significantly widen the gap between their own wealth and countries in Asia and Africa, with devastating consequences for colonized peoples (see for example Davis 2002). For example, in the 1920s a Belgian copper-mining company in Katanga, Zambia required family wages rather than individual wages in order to pay less (women and children's labor was 'incorporated' into the adult male labor wage; see Berger 1974). Colonizing nations' focus on manufacturing and mining accounted for much of the colonized nation's domestic product: in Taiwan, for instance, mining and manufacturing accounted for almost a quarter of the country's domestic product in 1938, similar to South Africa, Zimbabwe, and the Republic of the Congo (known as the Belgian Congo before Independence) in the 1960s (Austin 2014). The removal of private capital controls in the 1970s afforded greatly accelerated private capital flows in the form of direct foreign investment in capitalist enterprises in Asia, South America, and Africa. As Larry Neal and Jeffrey Williamson note:

> These private capital flows were on a scale that was orders of magnitude larger than were the aid flows from international agencies and government-to-government loans and grants in the early postwar decades. As we have seen in the years since, capitalist enterprises operating on a global scale can sustain repressive governments. This has often been the case with exploitable natural resources at locations that require mining or drilling operations. Large corporations exploiting government controls for natural resource extraction generally had the benefit of finance from impersonal equity and bond markets, starting with the English and Dutch East India Companies of the seventeenth century up to the multinational oil and mining companies of this century.
>
> (2014: 541)

Whilst colonies in tropical regions of the world attracted little private foreign investments into local economies, extractive industries invested heavily. For instance, between 1870 and 1937, an estimated 55.8 pounds sterling was invested in South Africa with its gold and diamond mines, compared with 3.3 pounds sterling investment in France's African colonies (Frankel 1938). According to Global-Data's mining database, the top three gold mining companies in South Africa in 2023 were Gold Fields (which operates the South Deep gold mine), Sibanye Stillwater (which operates the Kloof and

Driefontein gold mines), and Harmony Gold Mining (which operates the Mponeng and Tshepong gold mines) (Mining Technology
2023). As well as operating in South Africa, Gold Fields owns mines
in Australia, Ghana, Peru, Canada, and Chile. It amalgamated with
Gencor (owned by the Australian multinational company BHP)
in 1998. BHP, Gencor, and Gold Fields' origins are colonial: Cecil
Rhodes founded Gold Fields in 1887.

Sibanye Stillwater was a subsidiary of Gold Fields, founded in
2012, and which has had an equally multi-stage history of mergers
and acquisitions. Finally, Harmony Gold Mining operates in Papua
New Guinea as well as South Africa and their top shareholders
include the United Kingdom's Lingotto Investment Management,
LLP, M&G Plc, and Polunin Capital Partners Ltd, and the United
States' Van Eck Associates Corporation, Black Rock Inc., The
Vanguard Group, and Kopernik Global Investors LLC. Thus,
colonialism pushed capitalism's principle of individual land ownership in colonized regions, which remains highly controversial and
a primary cause of ongoing litigation in some countries (see Hird
and Predko 2024).

Mining produces different forms of waste, such as mine rock,
water pollution, tailings, dust emissions, and so on. As Mariette Lief-
ferink from the Federation for a Sustainable Environment notes:

> It is widely recognised that problems related to mining waste
> (tailings) may be rated as second only to global warming and
> stratospheric ozone depletion in terms of ecological risk. The
> release to the environment of mining waste can result in profound,
> generally irreversible destruction of ecosystems.
>
> (Mining Watch Canada 2022)

Chapter 1 details the fact that the resource extraction industry (oil,
gas, coal, minerals, metals) accounts for the largest portion of waste
globally. Much of this waste is produced at the mining sites themselves, and remains there during the life of the mine, and forever
afterwards. This means that whatever regions have or have had mining operations will also have mining waste legacies (Minchinton
1990; Ortiz 2020). For example, the Mine Tailings Working Group
of South Africa's *Tailings Report* notes that at 'the Witwatersrand gold
fields, over 120 mines have extracted over 116,000 tons of gold and

uranium, leaving a legacy of more than 270 tailings facilities that cover approximately 400 km^2' (Mining Watch Canada 2022). The report goes on to note that gold mine waste – creating 221 million tons or forty-seven percent of total waste – is the largest single source of waste and pollution in South Africa, and far greater than household municipal solid waste. Indeed, mining waste in South Africa accounts for *nearly half* of the country's total waste production.

CAPITALISM AND WASTE

The founding principle and aim of capitalism is to maximize profit. As such, waste is of primary concern to companies when they cannot extract profit from that excess. Manufacturers are continuously concerned with any overprocessing that does not add value to their 'bottom line' (that is, profit margin). Excess, in this context, is the constant threat of continuous growth, when surplus is tethered to profit. It is critical to note here that the crux of capitalist ruminations about excess are entirely concerned with its impact on profits: it does not concern itself with the impacts of material excess (waste) on human health or the environment (Bauman 2001). The resource extraction industry also considers mining waste to be a normal part of doing business.

But waste is also part of the capitalist logic. As the next section details, planned obsolescence occurs when businesses create products that break down by design. When these products break down, capitalism urges consumers to buy new products, thus creating waste and its negative health and environmental impacts. Capitalism also encourages consumers to buy new products through fashion cycles, where this year's bicycle is marketed as more desirable than last year's bicycle, and so on. Capitalism also produces waste as part of its logic through the intentional wasting of products when pricing shifts the profit advantage. For instance, in 2016, milk manufacturers wasted over forty-three million gallons worth of milk, dumping the milk in fields, manure lagoons, and animal feed (Liboiron 2016). More recently, farmers in Milwaukee and surrounding Midwest states have been pouring milk down sewers due to excess supply (Foreman 2023). As milk and cheese prices fell, capitalism encouraged farmers and industry lobbyists (such as Dairy Management Inc.) to make deals with fast-food companies such as Domino's Pizza and Taco Bell to include more dairy products on their menus (Gee 2016). In this way, at least some profit is made. But when no economically profitable options remain, capitalism incentivized farmers to reduce their cattle herds (by sending more cows to slaughter) and wasting the milk, rather than (for instance) partnering with the government to provide free milk to children in schools and/or providing free milk to income-insecure citizens.

Beyond the 'normal' (i.e. anticipated) amounts and kinds of wastes that resource extraction creates that have negative consequences for local human health and the environment, resource extraction creates waste of such scale, strength, impact, and speed that they are termed disasters. These disasters are the opposite of what Rob Nixon (2017) terms 'slow violence' whereby contamination (such as air pollution) affects certain communities over time, making it difficult to detect, identify sources and causes, and thus hold companies legally accountable. Between 1995 and mid-2020 there have been sixty-six major mine waste tailings dam failures worldwide – more than two every year (Pollon 2024). At the time of this writing, so far in 2024 alone, there have been several mining waste disasters. For example, on February 13th, the Çöpler gold mine in the Erzincan province of Turkey suffered a heap leach pad collapse, burying nine mine workers alive and causing an estimated ten million tons of cyanide-laced ore to pour into the valley below and towards the Euphrates River (Kneen 2024). Heap leaching is a process mining companies use to extract gold from crushed ore by spraying a cyanide solution over the rock that separates the gold from the rock. This particular heap leach pad was up to 257 meters high (the accepted maximum is 150 meters) and within 400 meters of a geologic fault line. On Februrary 20th, Bulla Loca, an illegal gold mine in La Paragua in Venezuela, collapsed, killing at least 14 people (Associated Press 2024). On the 6th of July, heavy rainfall caused a landslide in Sulawesi in the Bone Bolango region of Indonesia, killing at least ten people, including a four-year-old child and three women (Reuters 2024).

As Chapter 5 details, the world's largest and worst tailings dam disaster occurred in January 2019 when an iron ore mine in Brumadinho, Brazil killed at least 209 people and released over twelve million cubic meters of toxic sludge that destroyed communities. The gravity and scale of this one mining waste disaster was such that the International

Council on Mining and Metals (the largest mining industry group), the United Nations Environment Programme, and the Principles for Responsible Investment (pension investment) launched the Global Industry Standard for Tailings Management in 2020 as a set of best practices for mining waste management (see GlobalTailingsReview. org). As of February 2024, about half of all public mining companies have stated that they are committed to adopting the standard. The standard and its best practices are voluntary and include no mechanism to ensure compliance. In other words, it is not legislated, which would force companies to comply. The standard contains no ban on upstream dams, the cheapest kind of mine waste dam to construct, and the ones most vulnerable to leakage. And companies that do not rely on shareholder funding or are not concerned with public reputation have little incentive to comply (Pollon 2024). Nor does the standard apply to any of the 30,000 orphaned mines and abandoned tailings dams around the world. Moreover, the standard was set by the industry itself rather than an independent board, which raises serious credibility issues; or what is known as the 'who is watching the watcher' problem (Kemp and Hopkins 2021). Without very strong and binding regulations, mining disasters are predicted to increase in number and in severity as mining companies accelerate their resource extraction activities, including gleaning ever-increasing quantities of mining waste to recover increasingly scarce metals and minerals (Pollon 2024). According to Luke Fletcher, Director of the Australia Jubilee Research Centre in Sydney, 'there are no good environmental solutions. There are just different various versions of less bad' (in Pollon 2024).

The resource extraction industry also includes fossil fuels (oil, natural gas, coal). The Union of Concerned Scientists (2016) outline major externalities – the hidden costs of fossil fuels that the industry does not count in its public ledger – that include land degradation and pollution, as well as human health costs such as asthma and cancer. As Chapter 1 reviews, fossil fuels are extracted from the earth through either mining or drilling. As resources become more scarce, coal mining has shifted in such states as West Virginia, Virginia, Kentucky, and Tennessee in the United States from underground coal mining to surface mining in the form of either strip mining or mountaintop removal. Mountaintop removal requires that all of the mountaintop's topsoil and vegetation be removed. The non-governmental Appalachia advocacy group Appalachian Voices (n.d.) finds that mountaintop trees are often not even sold or used, but are simply burned or illegally dumped into valleys. When coal seams run deep beneath the surface, coal companies remove more of the mountain by using millions of pounds of explosives. Then the coal is dug up, and the excess waste is dumped into adjoining valleys. Then the burning of coal for energy produces carbon emissions and other waste that contribute to pollution and climate change. Underground coal mining causes considerable environmental impacts, including water and land contamination from acid mine drainage, fires, and the slow corruption of abandoned mines (Environmental Protection Agency 2015). Coal is known as a 'dirty fuel' because it contains toxic heavy metals and other chemicals. Sulfur-rich coal must be cleaned and refined before it is burned, leaving in its wake a coal slurry that contains arsenic, mercury, chromium, cadmium, and other heavy metals. According to the National Research Council (2010), up to fifty percent of pre-processed coal will end up as contaminated waste. After combustion, the coal ash left behind must be disposed of carefully as it contains toxic heavy metals. At least forty-two percent of the reservoirs in the United States that permanently store coal slurry and coal ash are unlined, and more are not properly maintained (Epstein et al. 2011). Oil and gas extraction creates its own waste, with heavy metals, hydrocarbons, and radioactive materials (National Research Council 2010). There is ample evidence that in the absence of tight regulations, the oil and gas industry does whatever it can to minimize the costs of waste disposal (Mall and Alemayehu 2021).

OUT-OF-SIGHT, OUT-OF-MIND: THROWAWAY SOCIETY

It is in private waste management's economic interests that society creates as much waste as possible. The burgeoning waste management industry encouraged the increase in waste production through various rationalities. As we have seen, growing concern for public health led to pressure on local governments to regulate waste collection, circulation, and disposal. Population growth was also a factor: more people were producing waste. And increasing production and consumption meant that this increasing population was producing more waste per capita. Besides increasing hygiene and health, throwing things away was rationalized as a time-saving activity. Single-use product advertising marketed time efficiency: far better to use single-use products than laboriously wash, mend, and find new uses for used products. Just as the motorized washing machine was marketed as time-saving, so was throwing away products after use. And not only was it apparently time-saving for working wives and mothers, it was advertised as far more convenient as well. Waste management companies were also hard at work convincing people that more materials were suited for disposal than for reuse. High on the list were food and clothing. Where once leftover food was creatively served to families after which it was given to domestic animals, and clothing was mended and re-sized for multiple family members, now companies encouraged households to get rid of the old and buy new; to put leftover food and unwanted clothing and other fabrics directly in the garbage bin.

As private companies contracted with municipalities, and waste was collected from the curb, people's connection with their waste – the time they spent dealing with their waste – significantly shortened. Waste management companies organized trucks to collect waste early in the morning, and encouraged an 'out-of-sight; out-of-mind' attitude. Waste management companies offered a solution to the extensive practice of dumping waste on land and in water systems. Companies began buying land outside of (upper- and middle-class) communities. They burned waste away from urban centers, investing in trucks and other infrastructure. Municipalities got on board, providing licenses and other permissions to dispose of waste near poor rather than wealthy communities (see Chapter 5).

Waste management companies also encouraged people to associate waste with social class: only poor people sorted through materials, reused, and made-do. People were encouraged by manufacturing, retail, and waste management companies to demonstrate their social class through wasting. The manufacturing, retail, and consumption revolution temporarily dipped during World War II (although war-products such as munitions, soldier uniforms and kits, and so on dramatically increased) and allied and enemy countries alike rapidly created public service campaigns to get people to gather and donate valuable materials in service to the war effort. While countries like Germany (East and West) and the United Kingdom maintained rationing for years after World War II ended, other countries – most notably the United States – witnessed a surge in consumption, and waste. New products replaced old products, and products for every part of the household, from small appliances to matching living room and bedroom sets, to the outfitted garage, to the surge in products for leisure activities (football, windsurfing, skiing etc.), were heavily advertised. Hobbies became consumer activities and power tools, card game paraphernalia, and 'how to' books became marketed investments to optimize leisure time (Strasser 1999). A whole industry of consumer magazines overtook the traditional Sears or other order-through-the-mail catalogues. All of this consumption was advertised to people as technologically superior, more convenient and hygienic, and a better reflection of how well one was doing in society. As the makeup company L'Oreal's 1971 advertisement proclaimed, 'You're worth it' (Specht and Erickson n.d.).

Chapter 3 details the consortium of American Can companies that replaced reusable glass bottles with plastic bottles by launching heavy media campaigns to shift people's attention from plastics waste to litter. Other companies employed the same tactics to introduce a number of products – such as disposable drinking cups and sanitary products – on to the market (Strasser 1999). Not only were these new products advertised as more hygienic (and therefore safer) but much more convenient. Much more convenient and time-saving, then, for the busy housewife to serve her family dried cereal than a cooked breakfast. Women in particular were promised a greater level of freedom from household drudgery with convenience products: the cereal box and leftover cereal tossed into the trash rather than the fat left over from breakfast painstakingly scraped into the fat jar kept cool, the other food leftovers re-served in yet more cooking, and the plates and cutlery washed at the sink. Consumerism, in short, promised (women in particular) freedom. Discarding things became an act of personal freedom: freedom to buy whatever we want, and freedom to toss it away when we no longer want it (Freidman 1962).

PLANNED OBSOLESCENCE

As Susan Strasser observes, 'economic growth during the twentieth century has been fueled by waste – the trash created by packaging and disposables and the constant technological and stylistic change that has made 'perfectly good' objects obsolete and created markets for replacements' (1999: 15). This is the basis of planned obsolescence: a policy that builds technological breakdown and aesthetic undesirability into products in order increase replacement purchasing. Planned obsolescence occurs when manufacturers advertise replacing working products (clothing, appliances, furniture, cars etc.) based on aesthetics. The light bulb is the first documented case of planned obsolescence (Rivera and Lallmahomed 2016). Even though filament technology in the 1920s produced bulbs that could last 2500 hours, producers cooperated to impose a 1000 hour limit on bulb lifetimes in order to increase the need for light bulbs, resulting in increased producer profit. Replacing functioning orange or green wall-to-wall carpeting in vogue during the 1970s with short-ply carpeting in vogue in the 1990s is an example of consumption for aesthetic reasons. It is designed to shorten the time span between purchases, known as the 'shortening of the replacement cycle' (Slade 2007).

Analyses of planned obsolescence often focus on car manufacturing. In the United States, up until the 1920s, Henry Ford had built his empire on reliable cars built with as few irregularities as possible from a streamlined assembly line. Ford's advertising platform was that his cars were highly dependable and designed to last. Ford was famously quoted as saying 'Any customer can have a car painted any color he [sic] wants so long as it is black' (Ford 1922/2008). This meant that Ford car owners had little need to buy new cars. In 1924, General Motors introduced yearly changes to their cars, heavily advertising style changes as desirable, and holding on to one's current car as undesirable. The advertising worked, General Motors sales outpaced Ford's and the planned obsolescence race to increase car sales began (Strasser 1999). In the decades since General Motors outsold Ford through constant product 'upgrading', planned obsolescence has gained enormous speed and traction, and has become an in-built feature of 'doing business' in our global capitalist society.

This is not to say that, along the way, planned obsolescence critics have not objected to this policy. Design engineers were highly skeptical of obsolescence caused by built-in product failure, and pointed out that this was anathema to the whole point of engineering. In his popular books, The Hidden Persuaders (1957), The Status Seekers (1959), and The Waste Makers (1960), Vance Packard extensively studied what he called 'obsolescence of quality' which referred to planned product failure, 'obsolescence of desirability', which referred to aesthetic changes in products designed to increase sales, and 'obsolescence of function', which referred to product modifications that decreased the product's performance. He demonstrated how manufacturers build in obsolescence to their products intentionally, to increase their profits. Packard pointed out that planned obsolescence depends on consumers not knowing how long a product can actually last. And he heavily criticized consumers for succumbing to the perpetually new 'shiny object'. According to Strasser (1999), Packard's arguments were sharply criticized by business, who argued that consumers wanted constantly updated products, and that 'out-with-the-old; in-with-the-new' was intrinsic to modern business.

It is beyond question that marketing is a significant contributor to increasing consumption. As Harrison Ford's character, Allie, from The Mosquito Coast put it:

> We eat when we're not hungry, drink when we're not thirsty. We buy what we don't need and throw away everything that's useful. Why sell a man what he wants? Sell him what he doesn't need. Pretend he's got eight legs and two stomachs and money to burn. It's wrong.
>
> (Weir 1986)

Since the early twentieth century, the list of single-use products has exponentially increased. Swanson's Hungry Man and other ready-made dinners in the 1950s (followed by ready-made lunches, breakfasts, and snacks) specifically advertised the disposability of packaging as an asset (Strasser 1999). Today, fashion and single-use plastics epitomize what François Jarrige and Thomas Le Roux refer to as the 'plastification of the world' (2020: 238). As Strasser notes:

> Packaging may have been disposable, but it was hardly worthless; the costs of packaging research and development, marketing, and raw materials were often equivalent to the cost of the contents. New materials and new combinations of paper, foil, and plastic went into food packaging, ranging from coated paper boxes to the three-section foil tray and six-color laminated box developed for the Swanson TV Dinner. These packages were fundamental

to marketing prepared foods… As the title of a 1967 American Management Association publication put it, *Packaging is Marketing*.
(1999: 268)

Planned obsolescence certainly made its mark in car manufacturing, as well as numerous other businesses. Two industries, fashion and fossil fuels, are particularly acute examples of businesses' use of advanced advertising and other techniques to encourage, shame, incite, and otherwise get people to buy their products.

TEXTILES

In contemporary affluent societies, textiles are closely linked with planned obsolescence. When people buy more clothes than they wear, we get fashion. And fashion has become one of the most significant contributors to climate change and plastics waste. The fashion industry is responsible for approximately ten percent of global greenhouse gas emissions (United Nations Economic Commission for Europe 2018). Clothing production is energy intensive, and includes very long supply chains, as fabrics are grown and synthesized in one or more countries, assembled into clothing in one or more other countries, and then shipped around the world to retailers. In fact, fashion production is more energy intensive than the aviation and shipping industries combined. In addition, the fashion industry uses a disproportionate amount of water: producing one kilogram of cotton required for one pair of jeans requires 10,000 liters of water, which is about the same amount of water one person drinks in ten years. And in turn, much of the water required to make fabric becomes waste water; about twenty percent of the global waste water production. Add to this the fact that about eighty-five percent of textiles end up in landfills, incinerators, or dumps. Keeping global temperatures below the 1.5 degree Celsius increase means tackling fashion production, circulation, and waste. Dyeing fabrics to produce the range of color choice we now have requires a host of synthetic chemicals, which also produces pollution as well as human health concerns:

Textile dyeing relies on chromium, lead, cadmium, sulfur, nitrates, chlorine compounds, arsenic, mercury, nickel and cobalt, formaldehyde-based dye fixing agents, chlorinated stain removers.

Dangerous on their own, these dye materials can also react with the disinfectant used to process fabric, particularly chlorine, and form by-products that are frequently carcinogenic. Dye effluent can clog the pores of the soil, crippling its productivity. Dumped into waterways, it can pollute drinking water and soil, impacting entire ecosystems, creating serious public health problems.

(Thanhauser 2023: 191–192)

Clothes washing produces a significant amount of microplastics: half a million tons per year (Abelvik-Lawson 2023).

Before the late nineteenth century – in the long 'waste-not; want-not' phase of our history – ready-to-wear clothing did not really exist. People (mainly women) made their clothes at home. Wealthy people wore clothes made by tailors and seamstresses. The main fabrics used were linen, cotton, and silk. Each fabric used significant amounts of organic material, water, soil, and labor to produce, but the smaller global population combined with the much less clothing that each person had meant that clothing had a much lower environmental footprint. Slowly, clothing became mass-produced. At first, undershirts and underwear were produced by poor pieceworkers in factories. 'Slop shops' that sold mass-produced clothing appeared in large US cities such as New York and Boston (Thanhauser 2023). The invention of the sewing machine and its wholesale purchase and use in factories greatly aided in the development of mass-produced fashion. As Sofi Thanhauser explains, mass-produced clothes were touted by industry as the solution to clothing slaves (with ready-made clothes slave labor could be directed to plantation and other work rather than making clothes), new immigrants, and soldiers. The American advertising industry, created in the 1920s, was pivotal in marketing mass-produced clothing as cheaper but of good quality, accessible, fashionable, and as exemplary of the American (and soon European) way of life. Mass-produced clothing advertisers, in other words, marketed clothing as antithetical to communism, and thus a key way to demonstrate allegiance to the 'free world' and capitalism.

Mass-produced clothing transformed again with the invention of synthetic materials. In what's known as the 'golden age of capitalism' chemical companies in tandem with university researchers dramatically increased fossil fuel consumption by creating synthetic fibers such as rayon and nylon, fertilizers and pesticides, and (single-use) plastics (Angus 2016). Scientists invented rayon, also called viscose, using pulped and liquified wood. It is typically blended with other materials like cotton to produce more fabric materials. For decades, workers making viscose fabrics experienced debilitating and sometimes lethal degenerative brain disease caused by the carbon disulfide used in the manufacturing process. Polyamide, more commonly known as nylon, is produced entirely from petroleum products, and was the brainchild of the DuPont chemical corporation. Nylon is used in an astonishing number of products, from clothing to parachutes.

It is estimated that the fashion industry produces over 1 billion items of clothing per year; over forty percent more than what can be bought or worn (Besser 2021). Affluent countries have such an excess of clothing that they export used (and even new) clothing to less affluent countries. According to the Ellen MacArthur Foundation, up to half of all the clothing donated to used clothing stores in developed countries ends up being exported to poorer countries (2017). These so-called 'donations' overwhelm these nations' domestic textiles and clothing industries, leading the East African Community (Tanzania, Burundi, Rwanda, Kenya, Uganda, and South Sudan) to ban second-hand clothing imports in 2019. And the influx of used clothing is so intense that some households are burning clothes to heat their homes and cook their food, releasing dangerous toxins into the atmosphere (Abelvik-Lawson 2023). Greenpeace also discovered that fashion industry companies also incinerate unsold clothes, for instance when clothing and accessories giant Burberry burned over 90 million pounds sterling worth of merchandise over a five-year period (BBC News 2018).

A recent study estimates that half of fast fashion is made out of plastics (Royal Society for Arts, Manufactures and Commerce 2021). Microfibers create microplastics water pollution because the microplastics are so small that they cannot be collected in washing machine or waste water treatment plant filters. Treated wastewater, containing microplastics, is released into rivers and oceans:

Microfibers represent the dominant source of plastic pollution in the ocean. They have been found in the sediments surrounding beaches, in mangrove groves, in Arctic ice, and in products for human consumption. 'The average person ingests over 5,800 particles of synthetic debris a year... and most of those particles are plastic fibers'. Every year, half a million tons of plastic microfibers seep into the ocean, the equivalent of fifty billion plastic bottles. Microplastics are a dramatic new problem for the world's water, even while older ones persist, now magnified by the sheer scale of garment production.

(Thanhauser 2023: 217–218)

Microplastics, as Chapter 1 details, effectively last forever in the environment.

FOSSIL FUELS AND PLASTICS

Throughout the twentieth century, fossil-fuel derived products dramatically increased, as did their manufacturing using fossil fuels as energy source. The fact that over half of the clothes we wear are derived from oil and gas is an indication of the pervasiveness of fossil fuels in contemporary global society: polyester, for instance, is derived from oil. Since the dawn of the industrial revolution, fossil fuels have become key to our global economic system. Researchers call this fossil capitalism (Altvater 2007). As fossil fuels – a non-renewable energy source – superseded renewable energy sources (wind, water, sunlight), energy use was no longer tethered to a place (beside a waterfall, for instance): instead, energy could be transported anywhere in the world (in oil barrels), stored and consumed twenty-four hours a day, 365 days per year (Graham 2021). The shift from steam to coal-generated electricity also allowed companies, such as mill operators in the United Kingdom in the nineteenth century, to control labor, relocating production to low-wage regions (see Chapter 5). This meant extended and dispersed supply chains, which in turn meant vastly increasing fossil-fuel burning by transport vehicles (trucks, ships, trains). And this meant more waste.

The concomitant increasing use of fossil-powered machines increased profits and further established fossil-fuel-based capitalism as the hegemonic global economic system. This also led to the manufacture of fossil-fuel-using (oil and gas) cars, and petroleum-based materials such as plastics, fertilizers, and pesticides. After World War

II, life became defined as 'homes, cars, yards' (Huber 2013: xiv). The massive expansion of fossil-powered transportation, from cars, to ships, to airplanes, and their associated infrastructure (superhighways, highways, roads, runways, and so on) exponentially increased waste production and pollution. And as Graham (2021) points out, the 'easy-access' and abundantly available oil and gas of the twentieth century has been eclipsed by peak-oil and far more unconventional fossil fuels in the form of Canada's tar sands, hydraulic fracturing, and deep-water drilling in the United States, the United Kingdom, Russia, and elsewhere, which evidence shows to be more environmentally detrimental than conventional oil and gas sources (Graham 2021; Grant et al. 2013; Howarth 2014; Hughes 2015; Ortiz 2020; see also Pineault 2018). And all of this creates more waste.

REVIEW

For most of human history, people regularly practiced reusing and recycling materials and products. Even manufacturers and retailers systematically reused and recycled materials as a money-saving practice. As consumer society advanced, recycling faded from mainstream view. When municipalities eventually contracted with private waste management companies these companies encouraged households to dispose more, as this increases waste management industry profits. Companies, in tandem, created more and more single- and short-use products, as well as sophisticated marketing campaigns to encourage consumption. This, as Chapter 3 details, has become part of the complicated contemporary ways through which we manage, and mismanage, waste.

SUGGESTED READING

Barnett, E. (2024). *Leftovers: A History of Food Waste and Preservation*. Bloomsbury Publishing.

Gille, Z. (2007). *From the Cult of Waste to the Trash Heap of History: The Politics of Waste in Socialist and Postsocialist Hungary*. Indiana University Press.

Strasser, S. (1999). *Waste and Want: A Social History of Trash*. Henry Holt and Company.

Walter-Toews, D. (2013). *The Origin of Faeces: What Excrement Tells Us About Evolution, Ecology, and a Sustainable Society*. ECW Press.

REFERENCES

Abelvik-Lawson, H. (2023). How Fast Fashion Fuels Climate Change, Plastic Pollution, and Violence, *Greenpeace*, 22 September. www.greenpeace.org/international/story/62308/how-fast-fashion-fuels-climate-change-plastic-pollution-and-violence/. Accessed 6 November 2023.

Altvater, E. (2007). The Social and Natural Environment of Fossil Capitalism, *Sociality Register*, *43* (*1*): 37–59.

Angus, I. (2016). *Facing the Anthropocene: Fossil Capitalism and the Crisis of the Earth System*. Monthly Review Press.

Appalachian Voices. (n.d.). *Mountaintop Removal 101*. https://appvoices.org/end-mountaintop-removal/mtr101/. Accessed 17 July 2024.

Associated Press. (2024). At Least 14 Confirmed Dead After an Illegal Open-Pit Gold Mine Collapses in Venezuela. *AP News*, 21 February. https://apnews.com/article/venezuela-illegal-mine-collapse-31d40848d930906d1f592117058e71b7. Accessed 17 July 2024.

Austin, G. (2014). Capitalism and the Colonies, in Neal, L. and Williamson, J.G. (eds.), *The Cambridge History of Capitalism*. Cambridge University Press.

Australian Research Council. (2022). A Shell Midden Packed with Thousands of Years of History Offers a Window into Australia's Past. www.arc.gov.au/news-publications/media/feature-articles/shell-midden-packed-thousands-years-history-offers-window-australias-past. Accessed 17 October 2023.

Bauman, Z. (2001). Excess: An Obituary. *Parralax* 7 (*1*): 85–91.

BBC News. (2018). Burberry Burns Bags, Clothes and Perfume Worth Millions. *BBC*, 19 July. www.bbc.com/news/business-44885983. Accessed 6 November 2023.

Berger, E.L. (1974). *Labour, Race and Colonial Rule: The Copperbelt from 1924 to Independence*. Oxford University Press.

Besser, L. (2021). Dead White Man's Clothes. *ABC News*, 21 October. www.abc.net.au/news/2021-08-12/fast-fashion-turning-parts-ghana-into-toxic-landfill/100358702. Accessed 6 November 2023.

Caves, R.W. (2004). *Encyclopedia of the City*. Routledge.

Davis, M. (2002). *Late Victorian Holocausts: El Niño Famines and the Making of the Third World*. Verso Books.

Ellen MacArthur Foundation. (2017). A New Textiles Economy: Redesigning Fashion's Future, 28 November. www.ellenmacarthurfoundation.org/publications/a-new-textiles-economy-redesigning-fashions-future. Accessed 7 December 2020.

Environmental Protection Agency. (2015). *Abandoned Mine Drainage*. www.epa.gov/nps/abandoned-mine-drainage. Accessed 17 July 2024.

Epstein, P.R., et al. (2011). Full Cost Accounting for the Life Cycle of Coal, *Annals of the New York Academy of Sciences*, *1219* (*1*): 73–98.

Fife, M. (2004). *The Nor Loch, Scotland's Lost Loch*. Scotforth Books.

Ford, H. (1922/2008). *My Life and Work*. CruGuru.

Foreman, P. (2023). Why US Farmers are Dumping Milk in Sewers, *Plant Based News*. 17 July. https://plantbasednews.org/culture/ethics/why-farmers-dumping-milk-sewers/#:~:text=Dairy%20farmers%20have%20long%20been,%2C%20environmental%2C%20and%20health%20concerns. Accessed 16 July 2024.

Frankel, S.H. (1938). *Capital Investment in Africa*. Oxford University Press.

Freidman, M. (1962). *Capitalism and Freedom*. University of Chicago Press.

Gee, K. (2016). America's Dairy Farmers Dump 43 Million Gallons of Excess Milk, *The Wall Street Journal*, 12 October. www.wsj.com/articles/americas-dairy-farmers-dump-43-million-gallons-of-excess-milk-1476284353. Accessed 16 July 2024.

GlobalTailingsReview. (2020). *Global Industry Standard on Tailings Management*. https://globaltailingsreview.org/wp-content/uploads/2020/08/global-industry-standard_EN.pdf. Accessed 17 July 2024.

Graham, N. (2021). *Forces of Production, Climate Change and Canadian Fossil Capitalism*. Haymarket Books.

Grant, J., Huot, M., Lemphers, N., Dyer, S., and Dow, M. (2013). *Beneath the Surface: A Review of Key Facts in the Oilsands Debate*. Pembina Institute.

Hird, M.J. and Predko, H. (2024) *Extracting Reconciliation: Inhuman Wastes, Indigenous Lands, and Colonial Reckoning*. Routledge Press.

Howarth, R.W. (2014). A Bridge to Nowhere: Methane Emissions and the Greenhous Gas Footprint of Natural Gas. *Energy Science and Engineering, 2* (2): 47–60.

Huber, M. (2013). *Lifeblood: Oil, Freedom, and the Forces of Capital*. Minneapolis, MN: University of Minnesota Press.

Hughes, D. (2015). *A Clear Look at BC LNG*. Canadian Centre for Policy Alternatives.

Jahan, S. and Mahmud, A.S. (2015). What Is Capitalism? Free Markets May Not Be Perfect but They Are Probably the Best Way to Organize an Economy, *Finance and Development*, June: 44–45. www.imf.org/external/pubs/ft/fandd/2015/06/basics.htm. Accessed 15 July 2024.

Jarrige, F. and Le Roux, T. (2020). *The Contamination of the Earth: A History of Pollutions in the Industrial Age*. The MIT Press.

Kemp, D. and Hopkins, A. (2021). *Credibility Crisis: Brumadinho and the Politics of Mining Industry Reform*. CCH Australia.

Kneen, J. (2024). Canada Turns Its Back on Mining Tragedy in Turkey, *Mining Watch Canada*, 23 February. www.miningwatch.ca/blog/2024/2/23/canada-turns-its-back-mining-tragedy-turkey Accessed 17 July 2024.

Liboiron, M. (2016). Dumping Milk from the Treadmill of Production, *Discard Studies*, 24 October. https://discardstudies.com/2016/10/24/dumping-milk-from-the-treadmill-of-production/. Accessed 16 July 2024.

Louis, G. (2004). A Historical Context of Municipal Solid Waste Management in the United States, *Waste Management Research*, 22 (*4*): 306–322.

Mall, A. and Alemayehu, B. (2021). *A Hot Fracking Mess: How Weak Regulations of Oil and Gas Production Leads to Radioactive Waste in Our Water, Air, and Communities*.www.nrdc.org/sites/default/files/fracking-mess-regulation-radioactive-waste-report.pdf Accessed 17 July 2024.

Mays, L. (ed.) (2010). *Ancient Water Technologies*. Springer.

Minchinton, W. (1990). The Rise and Fall of the British Coal Industry: A Review Article, *VSWG: Vierteljahrschrift für Sozial- und Wirtschaftsgeschichte*, 77: 212–226.

Mining Technology. (2023). *The Five Largest Gold Mines in Operation in South Africa*. www.mining-technology.com/marketdata/five-largest-gold-mines-south-africa/?cf-view. Accessed 15 July 2024.

Mining Watch Canada. (2022). *Mine Waste Leaves Toxic Legacies in South Africa – And International Launch of Tailings Report*. https://miningwatch.ca/news/2022/5/24/mine-waste-leaves-toxic-legacies-south-africa-and-international-launch-tailings. Accessed 15 July 2024.

National Research Council. (2010). *Hidden Costs of Energy: Unpriced Consequences of Energy Production and Use*. The National Academies Press. www.nap.edu/catalog/12794/hidden-costs-of-energy-unpriced-consequences-of-energy-production-and. Accessed 17 July 2024.

Neal, L. and Williamson, G. (2014). The Future of Capitol, in L. Neal and J.G. Williamson (eds.), *The Cambridge History of Capitalism*. Cambridge University Press.

Nixon, R. (2018). *Slow Violence and the Environmentalism of the Poor*. Harvard University Press.

Ortiz, R.J. (2020). Oil-Fueled Accumulation in Late Capitalism: Energy, Uneven Development, and Climate Crisis, *Critical Historical Studies*, 7 (*2*): 205–240.

Packard, V. (1957) *The Hidden Persuaders*. IG Publishing.

Packard, V. (1959) *The Status Seekers*. David McKay Co.

Packard, V. (1960). *The Waste Makers*. IG Publishing.

Pineault, E. (2018). The Capitalist Pressure to Extract: The Ecological and Political Economy of Extreme Oil in Canada. *Studies in Political Economy*, 99 (*2*): 130–150.

Pollon, C. (2024). It's Never Been More Important to Find a Mining Waste Fix, *National Observer*, 8 February. www.nationalobserver.com/2024/02/08/analysis/Brumadinho-Brazil-Polley-mining-disasters-tailings-dam-breach-safety. Accessed 17 July 2024.

Reuters. (2024). Landslides Kill 10 on Indonesia's Sulawesi Island, 40 Missing, *Reuters*, 8 July. www.reuters.com/world/asia-pacific/landslides-kill-12-indonesias-sulawesi-island-18-missing-2024-07-08/. Accessed 17 July 2024.

Rivera, J.L. and Lallmahomed, A. (2016) Environmental Implications of Planned Obsolescence and Product Lifetime: A Literature Review. *International Journal of Sustainable Engineering*, 9 (2): 119–129.

Royal Society for Arts, Manufactures and Commerce. (2021). *Half of 'Fast Fashion' Entirely Made of New Plastics.* www.thersa.org/press/releases/2021/half-of-fast-fashion-entirely-made-of-new-plastics-study. Accessed 5 November 2023.

Slade, G. (2007). *Made to Break: Technology and Obsolescence in America.* Harvard University Press.

Specht, I. and Erickson, M. (n.d.). 15: L'Oréal (1971) – Because I'm Worth It, *Creative Review.* www.creativereview.co.uk/because-im-worth-it-loreal/. Accessed 31 October 2023.

Strasser, S. (1999). *Waste and Want: A Social History of Trash.* Henry Holt and Company.

Thanhauser, S. (2023). *Worn: A People's History of Clothing.* Vintage Books.

Tulchinsky, T. (2018) John Snow, Cholera, the Broad Street Pump; Waterborne Diseases Then and Now, *Case Studies in Public Health*, 77–99. https://doi.org/10.1016/B978-0-12-804571-8.00017-2

Union of Concerned Scientists. (2016). *The Hidden Costs of Fossil Fuels.* www.ucsusa.org/resources/hidden-costs-fossil-fuels. Accessed 17 July 2024.

United Nations Economic Commission for Europe (UNECE). (2018). UN Alliance Aims to Put Fashion on Path to Sustainability. 12 July. https://unece.org/forestry/press/un-alliance-aims-put-fashion-path-sustainability. Accessed 25 June 2024.

Velis, C.A., Wilson, D.C., and Cheeseman, C.R. (2009). 19th Century London Dust-Yards: A Case Study in Closed-loop Resource Efficiency, *Waste Management*, 29 (4): 1282–1290.

Waste Management Inc. (n.d.). *Our Story.* www.wm.com/ca/en/inside-wm/our-story. Accessed 4 November 2023.

Weir, P. (dir.) (1986). *The Mosquito Coast.* P. Theroux and P. Schrader (writers). The Saul Zaentz Company.

UNDERSTANDING WASTE

OVERVIEW

Chapter 3 considers how different understandings of waste determine how waste is managed. Waste management companies understand waste as a commodity. Local governments understand waste as something their constituents expect will be managed with as little tax money as possible. Manufacturing companies understand waste as inevitable to the product production process. The energy sector increasingly understands waste as an alternate source of energy, and therefore source of profit and a means of achieving net-zero carbon emissions commitments. And everyday citizens largely define waste as something that they should take responsibility for reducing through more and better individual and household recycling.

INTRODUCTION

Waste is a multi-faceted complex global problem. Arguably, everyone is affected by waste. Everyone who directly handles waste, from extremely low paid waste pickers to waste management employees to individuals and households sorting their waste into disposal and recycling, understands waste in particular ways (as source of employment, individual responsibility, daily chore and so on). People in middle-income and wealthy regions understand waste as something

DOI: 10.4324/9781003398424-3

their municipalities are responsible for managing through contracts with private waste management companies paid for through taxes. Waste management companies and corporations understand waste as a commodity through which its management accrues exponentially increasing economic profit.

These understandings come from our direct and indirect (such as watching documentaries or viewing posts on social media) experiences. And they are also significantly influenced by waste stakeholders, which are overwhelmingly made up of industries that profit from waste, and therefore have an enormous stake in steering narratives about waste in particular directions. As Chapter 4 details, the plastics industry has effectively guided government and public discussions about marine plastics waste towards understanding ocean plastics as an individual and community responsibility, and chemical and mechanical plastics recycling as key to meeting plastics reduction goals. Numerous beach clean-up events are sponsored by plastics corporation conglomerates, and plastics recycling appears to offer a viable solution to increasing plastics production, an example we return to later in this chapter.

WASTE STAKEHOLDERS

Waste is a global problem whose lived experience varies substantially depending on cultural history, nationality, social class, gender, race and other structural factors. For instance, people living in privileged circumstances (income-secure, homeowning, non-racialized) may frame waste as a problem of individuals not 'doing their part' to sort through their waste to decrease waste and increase recycling. Dispossessed people living in poverty might frame waste as an injustice whose remedy is affluent countries accepting financial and ethical responsibility for the mountains of waste they daily export to impoverished countries.

The major stakeholders who differentially understand and define waste issues are: (1) extraction and manufacturing industries; (2) experts (who may or may not work for industry or government); (3) local, regional, national and international government agencies; (4) non-governmental organizations and social media; and (5) various publics who may or may not organize in communities. Each stakeholder frames waste issues according to their own interests. Rightholders, as Chapter 5 more closely examines, are people on whose traditional lands colonizing forces have settled or otherwise extracted resources from, producing waste legacies in the process. From the Saami in northern Sweden to Inuit in northern Québec, Kanak peoples in Nouvelle Calédonie to the Jarunda people of the Brazilian Amazon, Rightsholders claim rights and stewardship of lands and waters, some of which have (and are) literally waste dumping grounds of successive industries, governments and colonizing countries. Rightsholders are increasingly framing waste as a social injustice issue.

UNDERSTANDING WHAT THE PROBLEM OF WASTE IS

Understanding issues in particular ways is called framing, and it is a powerful means by which we make sense of the world. As Robert Entman notes, framing is the means by which individuals 'select some aspects of a perceived reality and make them more salient...in such a way as to promote a particular problem definition, causal interpretation, moral evaluation, and/or treatment recommendation' (1993: 51). In contemporary society, the media provides an excellent example of framing.

Private companies and interest groups are well aware of the power of framing. PlasticsEurope, Coca-Cola, ExxonMobil and other companies use framing to encourage the public to focus on the positive aspects of their products (such as convenience) rather than the human and environmental harms that their products cause (such as pollution and contamination). The dominant understanding of waste is that more individuals need to take responsibility for reducing waste through recycling which, coupled with innovations in disposal and recycling technologies, will resolve the global waste crisis. Industry and governments purposefully encourage people (through advertising and other means – see Chapter 4) to understand waste in this way in order to draw attention away from their disproportionate contribution to the waste crisis. Waste is understood in one of three major ways, according to which stakeholder is framing the issue: individual responsibility, economic resource, or social injustice.

UNDERSTANDING WASTE AS INDIVIDUAL RESPONSIBILITY

Manufacturing and retail industries, local and national governments, numerous volunteer organizations, and many members of the public present waste as an individual problem whose solution is greater individual responsibility. Simply put, they define the problem as individuals and families producing too much waste. The solution is, therefore, for individuals and families to decrease the amount of waste they are producing, primarily through recycling. To some degree, this frame relies on 'common sense' insofar as it concentrates on the garbage that most of us see and handle on a daily basis: municipal solid waste (MSW; see Chapter 1 for definition). It also reinforces a predominant axiom within western cultures that prioritizes both individual freedom and individual responsibility. This axiom is expressed in common sayings such as 'the early bird gets the worm' and 'you reap what you sow'.

As an example, at the university where I work, there are recycling and garbage bins in all of the buildings. Above each bin is a photograph of examples of things that are appropriate to place in each bin: a photograph of a banana peel above the compost bin; a photograph of a plastic water bottle above the plastics recycling bin; a photograph of a piece of paper above the paper and cardboard recycling bin; and a potato chip package above the garbage bin. Above these photographs is written in large bold letters, 'One mistake makes the entire bin *garbage*. Please sort your recycling properly'. There are many messages such as these, to act responsibly by managing our waste (in this case, to sort the things we are discarding in the correct bin). Taking responsibility is part of a wider moral and ideological principle called individualism. Individualism focuses on the individual – their interests and values, and their self-reliance and responsibility for themselves. Individualism emphasizes the individual person's responsibility for themselves as opposed to the government or state's responsibility for individuals. Individualism is critical to neoliberalism as a way of governing populations.

Waste management, as individuals mainly encounter it, almost exclusively concerns dealing with waste once a product has been consumed, or post-consumption waste. For instance, electronic waste discussions almost always focus on post-consumption: all of the iPhones, Android phones, iPads, laptops, ear buds and desktop computers individual consumers no longer want. We focus much less on all of the waste that is produced in the resource extraction and manufacturing of these and other electronic products. In fact, most people have no idea how much waste the manufacturing of any given product creates. And it isn't easy to find this information:

manufacturers are largely not legally required to disclose how much waste they produce in manufacturing their products. In *How Bad Are Bananas? The Carbon Footprint of Everything* (2011), Mike Berners-Lee calculates how much carbon different products, events and services use. A pair of Crocs shoes, for instance, made of 250 grams of expanded ethylene-vinyl acetate, and sold without packaging, creates about 1.5 kilograms of carbon dioxide. Leather shoes, by contrast, produce about fifteen kilograms of carbon dioxide. About fifty percent of this carbon footprint, Lee estimates, comes from the materials, around twenty-five percent is produced in manufacturing the shoe, fifteen percent in transporting the shoe from manufacturer to consumer (more if it is by plane rather than ship), and another five percent for miscellaneous things. Leather shoes also produce more carbon dioxide due to the carbon intensity of cattle farming. Carbon dioxide is just one waste that manufacturing shoes creates. There are also chemicals in dyes, glues, plastics and so on.

Industries extract and consume an extraordinary array and volume of materials in fabricating consumer products. For example, about ninety million pounds of e-waste is recycled through Dell's Reconnect program in the United States. This is a significant amount of electronics that would otherwise go to disposal. But a single *smelter* operation in Mexico that processes the copper and other metals required to manufacture electronics produces 819,000 million tons of sulfuric acid waste (Lepawsky 2018). Thus, the waste produced in processing just some of the material used in manufacturing electronics from one smelter is 1.8 times larger than e-waste exports from the entire United States in a year. The case of e-waste is generalizable to many extraction and manufacturing sectors including oil and gas (that produces plastics), agriculture (that produces meat, poultry and other organics), mining (that produces electronics, including cars), and so on.

One of the little-known realities of our global waste problem is that most waste is generated at the extraction and production stages rather than at the consumer stage of a product's life. In Canada, for instance, less than two percent of the total waste produced is MSW (Hird 2021). In the United Kingdom, MSW accounts for approximately twelve percent of the country's total waste (Government of UK 2023). Silpa Kaza et al. (2018) found that the amount of waste that industries alone produce is eighteen times higher than post-consumption waste.

Moreover, a closer look at the contents of MSW reveals that most of it (over 40 percent) is packaging: 77.9 million tons generated in the United States alone in 2015, and almost thirty percent of the US's total waste generation (US Environmental Protection Agency, 2022). Packaging is primarily made out of plastics, aluminum, paper, cardboard, wood, glass and other materials. Individuals and families are responsible for sorting through packaging with their unpaid labor. Waste and recycling companies then charge municipalities to collect these items, which municipalities pay for through taxes. Waste and recycling companies then sell these items for profit, which they do not share with consumers. In other words, consumers sort through waste for free; companies sort through waste for profit (Liboiron 2014).

And yet, most people assume that our waste problems stem from individual consumption and not enough recycling. And the reason people believe this is that manufacturing industries, waste management companies, and governments purposefully present waste in ways that almost exclusively focus the public's attention on post-consumption waste and the public's responsibility for this waste. In this way, as Max Liboiron points out, "the individual rather than government or industry is represented as the primary unit of social change" (2010: 1). The global neoliberal capitalist political economic structure depends on the acceptance of this understanding because it is driven by increasing circuits of resource extraction (primarily oil, gas and coal but increasingly the 'critical minerals' required for electric cars, electronics and so on), manufacturing products, and distributing all of these products around the globe. The waste management industry itself is part of this political economy of constant growth: just dealing with MSW went from $205 billion in 2010 to a projected $375 billion by 2025 (Wilson and Velis 2015).

The Keep America Beautiful 'Crying Indian' advertising campaign in the United States is a good illustration of the manufacturing industry purposefully presenting waste as the public's problem and thereby deflecting attention away from their own responsibility in creating waste. In the 1950s, beverage companies such as Coca-Cola were moving from glass to plastic pop bottles. The bottle deposit systems required the companies to buy-back the bottles, clean and redistribute them. The buy-back system was better for the public and for the environment, but garnered less profit for the bottling companies (MacBride 2012; Rogers 2006). Plastic bottles were easier and cheaper to manufacture, thereby increasing company profit.

In 1953, the state of Vermont tried to institute a compulsory bottle deposit that consumers would pay when they purchased bottles and a ban on selling beer in non-refillable bottles. In response, a consortium of American beverage companies, including the American Can Company, Continental Can Company, the US Brewers Foundation and the Owens-Illinois Glass Company, quickly responded by introducing the Keep America Beautiful campaign, which focused on the problem of highway litter. Government agencies such as the Connecticut State Highway Department and the New York State Department of Public Works, as well as non-profit organizations such as the Izaak Walton League of America, quickly joined in the deflection effort. Together, the bottling industry organized and paid for brochures that were widely distributed via post to people's homes as well as several public service announcements designed to educate the public about the problem of public littering.

In 1971, coinciding with the world's second Earth Day, the Keep America Beautiful campaign launched a television advertisement with the tag-line 'People Start Pollution. People Can Stop It', which became known as the 'Crying Indian' advertisement. The advertisement features a pretend American Native Indian, Iron Eyes Cody, paddling a birch-bark canoe down a river. As he paddles his canoe to shore and steps out of the canoe, a white woman throws trash out of her car, which lands at Iron Eyes Cody's feet. As Cody looks at the trash, a single tear falls down his cheek. At this time of writing, this advertisement is still available to watch on YouTube. This, and the American Chemistry Council's Plastics Make it Possible campaign, coined the term 'litterbugs' to focus on individuals who discard waste in public spaces.

The Crying Indian advertisement is an excellent illustration of industry and government successfully framing waste as a post-consumer (litter) problem created by individuals (members of the public discarding their individual waste on the roadside) and away from pre-consumer waste produced by manufacturing and retail industries whose profits increase when they are not regulated to provide, and pay for, bottle-refilling (reuse) services. The persuasive advertising campaign offered a simple message contrasting nature's purity against people's carelessness and disregard for the environment (for more details, listen to NPR's 2019 podcast, The Litter Myth. The success of the Keep America Beautiful campaign in presenting the public as the problem rather than industry eventually led the tobacco industry to fund campaigns that carefully diverted attention away from the human health and environmental risks of smoking (which would lead to a decrease in tobacco industry profits) and towards the problem of cigarette-butt littering. The Crying Indian and other Keep America Beautiful campaigns are regarded as the first highly successful example of corporate greenwashing. Corporate greenwashing occurs when companies and corporations exaggerate their green credentials, or the environmental good they do (Watson 2016).

Facing what can be overwhelming images and news stories about mounting waste and pollution in the world's waterways, on land, in the atmosphere, and even inside the bodies of humans and animals, it is understandable that people are seeking ways to help with this global catastrophe. One increasingly popular activity that more and more people (especially people living near coastlines, and who have the financial security and time security to engage in volunteer work) are joining beach cleanups. September 16th has been declared International Coastal Clean-Up Day.

The Ocean Blue Project is one such initiative. According to its website, the Ocean Blue Project was founded in 2012 in Newport, Oregon, by brothers Richard and Fleet Arterbury, to help clean up plastics waste in water. The Project claims to have removed over 1.25 million pounds of ocean plastics through the efforts of thousands of volunteers across the United States. Their work includes: 'plastic disaster relief projects' that remove plastics from water with the goal of recycling the removed material; 'community-led beach and river ocean cleanup projects'; school curriculum projects that teach Kindergarten to high school students about 'environmental stewardship', and research partnerships that support research projects that 'investigate practical local solutions to pollution problems' (Ocean Blue Project n.d.).

The first three initiatives directly involve volunteers (i.e. unpaid labour) in either physically removing plastics waste from waterways or in teaching children and adolescents how to do this. The fourth initiative funds projects concerned with developing technological ways of dealing with plastics in water such as filtration systems to capture plastics before they enter water systems and systems that break down plastics. These research projects, too, involve volunteers to install and test the technologies.

Coastal clean-ups such as The Ocean Blue Project and The Ocean Cleanup are illustrative of promoting individual responsibility because not only do thousands of people world-wide participate in this (non-paid) activity because it is understood as an effective solution, but the activity shifts attention away from the source of the problem. As Pichmol Rugrod observes, 'it's future work and highlights how plastic pollution is more than a waste management problem' (2023: np). The Ocean Cleanup receives funding from Coca-Cola and the Saudi Aramco oil company, and as such is criticized for actually helping plastics companies in their greenwashing efforts (Walsh 2023). The Ocean Cleanup, like The Ocean Blue Project and other cleanup organizations, focus on cleanup technologies: technologies that are entirely directed to post-consumption plastics rather than plastics production (see, for example Kakadellis and Rosetto, 2021, for a review of technologies towards a 'circular bioeconomy'). To foreshadow the next section, these technologies also serve the 'waste as resource' understanding that industry promotes.

For example, the International Coastal Cleanup and Ocean Conservatory report, *Stemming the Tide* (2015), called for mass incineration infrastructure to be built in Asia by United States and European waste management companies to deal with marine pollution. As Max Liboiron (2015) observes, the report's steering committee included Coca-Cola, Dow Chemical, the America Chemistry Council and the World Wildlife Fund. In response, the Global Alliance for Incinerator Alternatives (GAIA 2015) wrote an open letter responding to the report. Signed by over 800 grassroots organizations and individuals from over ninety countries, the open letter points out that not only does incineration only deal with post-consumption plastics (and therefore does not in any way stem the tide of plastics waste, but indeed encourages it since there is a proclaimed 'solution') but that it does nothing to reduce our reliance on fossil-fuel derived plastics. In 2022, the Ocean Conservatory finally released a 'Stemming the Tide Statement of Accountability', in which the organization 'unequivocally rescind[ed] any direct or indirect endorsement of incineration as a solution to ocean plastic pollution' (2022). The statement further reads:

> In *Stemming the Tide*, Ocean Conservancy focused solely on minimizing the amount of plastics entering the ocean. We investigated and included incineration and waste-to-energy as acceptable solutions to the ocean plastic crisis, which was wrong. We failed to confront the root causes of plastic waste or incorporate the effects on the communities and NGOs working on the ground in the places most impacted by plastic pollution. We did not consider how these technologies support continued demand for plastic production and hamper the move to a circular economy and a zero-carbon future. Further, by focusing so narrowly on one region of the world (East and Southeast Asia), we created a narrative about who is responsible for the ocean plastic pollution crisis — one that failed to acknowledge the outsized role that developed countries, especially the United States, have played and continue to play in generating and exporting plastic waste to this very region. This too was wrong.

Using the metaphor of dealing with an overflowing bath tub, Katharine Owens and Katie Conlon argue that nets designed to scoop plastics waste from waterways and attempts to purify waterways by using bacteria and other organisms to absorb or breakdown plastics are ways of 'mopping up' water from the bathroom floor (2021: 1). As good as this mopping up might be, it does nothing to turn off the tap producing the flood. Or as Chistina Dixon from the Environmental Investigation puts it, 'it's like sticking a Band Aid on a broken leg' (in Walsh 2023: np). These metaphors point to the difference between dealing with the after-effects of a problem and dealing with the source of the problem itself.

This is not to criticize beach cleanup volunteers, or indeed anyone who volunteers their time and energy in any post-consumption waste management activity. The point is to understand that these individual actions do nothing to prevent waste production. And without simultaneously and explicitly protesting against the industries that are producing waste, individual actions actually serve the interests of industry and government in deflecting attention away from the real culprits. Industries, of course, do not want any public or government attention on them as waste producers. Governments within systems that make them vulnerable to industry lobbying do not want to challenge these industries. The result is that industries and governments often work in (implicit) tandem to encourage individual responsibility for waste.

Some organizations are keenly aware that their grassroots volunteer-led community actions to clean up beaches are only part of their mission: they have producer responsibility firmly in their cross-hairs and use their work to focus on 'turning off the tap'. For example, Greenpeace Thailand cleans up plastics litter on Thailand's beaches. At the same time, it conducts plastics brand audits on various beaches. The results of these audits, conducted between 2018 and 2022, show that most of the plastics waste is produced by five local brands (Charoen Pokphand Group, Dutch Mill, Osotspa, Sermsuk, and Singha Corporation) and 5 international brands (Coca-Cola, PepsiCo, Nestlé, Unilever and AJE Thai) (Rugrod 2023). This and similar audits (for example, see King 2017 about plastics waste audits in the Philippines) bring attention to the source of ocean plastics waste: big business, and put pressure on these companies to take meaningful steps to reduce the plastics waste they produce. They help social movements such as #Break-FreeFromPlastic by providing valuable audit data that exposes the companies who are responsible for mounting plastics production and its waste. Stopping the source of the literal tide of ocean plastics has a much greater and long-lasting impact on waste reduction than does cleaning up sections of beaches only to have them saturated with plastics within hours. More generally, it is clear that citizen's real power is in motivating government regulations and policies that restrict and otherwise limit plastics production (see Chow et al 2017 and Schnurr et al 2018).

The Crying Indian and Keep America Beautiful campaigns (see above), beach cleanups and the like are part of what Nusa Urbancic and her colleagues at the Changing Markets Foundation (2020) call 'tactics to delay, distract and derail' that corporate plastic polluters use to avoid mandatory measures to reduce plastics waste. In a sweeping study that focused on the 10 largest plastics polluters – including Coca-Cola, Colgate-Palmolive, Danone, Mars Incorporated, Mondelez International, Nestlé, PepsiCo, Perfetti Van Melle, Procter & Gamble, and Unilever – the researchers assessed these companies': support for progressive legislation; plastics reduction target ambitions; commitment to reuse; recycled content, including substituting plastic with other single-use materials; transparency and accountability, which includes to what degree their commitments are applied across all markets they operate within or only some; and to what extent their solutions to the plastics crisis depends on false solutions such as chemical recycling, mechanical recycling (see Chapter 4) and what is known in the industry as 'lightweighting'. Lightweighting is the redesign of plastic packaging to be thinner, thus using less plastics. Thinner plastics packaging does not affect the environmental consequences of the item, and may also negatively impact its ability to be recycled.

The analysis brings into sharp relief that not only do these companies rely heavily on the public's individual responsibility as a tactic to delay and distract from their responsibility in producing plastics waste, but actually rely on the volunteerism of setting and implementing non-binding targets for plastics reduction. There is little or no enforcement to these targets, and companies are largely left to voluntarily report and adhere to the targets they themselves make. In this way, plastics polluting companies add insult to injury when they heavily deploy individual responsibility.

Companies spend millions of dollars annually on public relations, advertising, 'educational' programs and volunteer associations all of which are designed to blame the consumer for littering and not sorting their trash properly. In 2019, Coca-Cola, the world's biggest plastics waste polluter, launched the 'Don't buy Coca-Cola if you're not going to help us recycle!' campaign with billboards in Belgium and other countries (Arthur 2019). And it blamed consumers for its continued production of plastics, stating that customers prefer plastic bottles (Bandoim 2020). The company's justification does not mention that people around the world are deeply concerned with plastics (and particularly single-use) pollution and are calling on governments to enact far stricter regulations requiring companies to switch (back to) refillable bottles. Nor does it acknowledge the large and sustained consumer opposition when it initially converted from reusable glass bottles to single-use plastics bottles.

This played out during the Covid-19 global pandemic, which plastics producers used to justify their business-as-usual model. In fact, the American Chemical Council and Novolex paid for research studies that claimed that reusable grocery bags spread the virus (see Johansson 2021). In 1990, Ronald Bruner from The Society of the Plastics Industry, Inc. extolled the virtues of the plastics industry in working with US government administration 'to implement public service educational campaigns with specific messages targeted to selected audiences' (1990: 1077). The industry-supported (Sustainable Packaging Coalition) organization How2Recyle website states: 'End users of packaged goods – citizens – are integral to sustainable material management. Without their participation the material loop cannot be properly closed' (in Urbancic et al., 2020: 90). There is a disturbing social injustice element to this citizen-blaming, which we re-visit in Chapters 5 and 6:

> Instead of taking responsibility for the waste it creates, the industry now blames low- and middle-income countries – especially in Asia – for ocean plastics, due to their 'lack of waste-management infrastructure'. In doing this, they are conveniently forgetting that most ocean plastic pollution consists of the products these corporations sell there, and that many of these countries also face the double burden of being the dumping ground for the world's exported waste.

The biggest plastics polluters also organize together to form organizations with solution-oriented names such as the Alliance to End Plastic Waste, the Trash Free Seas Alliance, the Plastics Recycling Foundation, the Council for Solid Waste Solution, Closed Loop Partners, the RP, Ecoembes, the Sustainable Packaging Coalition, and the Global Plastic Action Partnership. By investing in these organizations, polluting companies make a front-facing commitment to solving the problem they have created, but the lack of transparency, no requirements to meet targets, and the heavy focus of these organizations on individual volunteerism (such as beach cleanups) means that the organizations themselves are part of the problem. Indeed, there is a 'shocking amount of overlap between corporate membership of the initiatives that claim to solve plastic pollution and trade associations and lobby groups that actively work to undermine ambitious legislation' (Urbancic et al. 2020: 15).

Companies such as Coca-Cola and PepsiCo that produce single-use plastics are not alone in encouraging individual responsibility as an effective deflection away from their own responsibilities for waste, pollution, and climate change. Recent research demonstrates that Exxon, Mobil and ExxonMobil used this same framing strategy to publicly amplify consumers' contribution to climate change and deflect attention away from their own, much greater, oil and gas industry contribution, as well as using paid editorial-style advertisements to undermine climate change science (and scientists) while acknowledging the climate change reality in their internal documents (Supran and Oreskes 2020). Indeed, 'in private and academic circles since the late 1970s and early 1980s, ExxonMobil predicted global warming correctly and skillfully' while spending millions of dollars from the early 1970s to convince people that fossil fuels do not significantly contribute to global warming (Supran, Rahmstore and Oreskes 2023: 1). These case studies are detailed in Chapter 6. At this time of writing, California Attorney General Rob Banta is suing ExxonMobil for 'deceiving the public about the recyclability of plastics' (State of California Department of Justice 2024).

UNDERSTANDING WASTE AS AN ECONOMIC RESOURCE

In 1993, Harold Crooks wrote that 'trash is our only growing resource' (22). Crooks was calling attention to the irony that while we are massively decreasing the Earth's natural resources, we are greatly increasing our waste generation. What he could not have known then is the rapidly growing drive to literally transform waste itself into a material and economic resource.

In 2005, Alexander Judd wrote *In Defense of Garbage*, in which he argued that waste should be defined as any land that is *not* being used for landfilling garbage:

> The public perceives that the garbage crisis is caused by the run-away growth of disposables, packaging, and discards in general. The real problem, of course, is not the growth of garbage or the quantity of garbage; it is the closing of landfills and the failure to provide replacement sites or alternate ways to handle the discards of towns and cities. The production of garbage responds to growth in population, household formations, affluence, and commercial activity, but the capacity for the disposal of waste depends more on the availability of land – space – than any other factor.
>
> (2005: 21)

For Judd, the problem is not increasing waste: it is wasting unused land that could be put to use in order to bury and store more waste. The view that everything is a potential resource waiting to be realized through human technical and entrepreneurial ingenuity is a common way of understanding the global waste crisis. Industry and governments around the world increasingly argue that waste is a valuable *resource* that may be put to good use. Indeed, within some industries and government agencies, waste is (definitionally) disappearing and is being replaced with profitable resource.

Extractive and manufacturing industries, experts and governments are increasingly framing waste as something that is not waste at all, but a valuable (i.e. profitable) material resource. For instance, Laurence Rocher (2020) analyses how waste is increasingly framed as a valuable energy-producing resource, or as Rocher puts it, 'waste has become an energy issue' (2020: 98). Using France's energy sector as a case study, Rocher's research shows how plastics waste is defined as something that is useless to the public (i.e. waste), which municipalities collect (with the costs externalized to taxpayers, and using the unpaid labour of citizens), and which companies transform into energy (electricity) to sell back to customers. Waste as resource is thus defined as a source of profit (it makes money) and a public good (it provides a service – electricity that supplies households).

This is how, for instance, Sweden is able to boast that it has resolved its waste problem. Headlines such as 'Sweden's strange problem: not enough trash' (Murphy 2017) present Sweden as a country of almost 'zero trash'. The Swedish government's own 'The Swedish Recycling Revolution' (Hinde 2020; see also Savini 2019) claim is based entirely on the reframing of waste as a source of energy (in Sweden's case, electricity used for heating). What this understanding of waste as a resource does not detail is the environmental costs of waste-to-energy incineration facilities, including highly toxic waste generation and the significant environmental harm accrued by transporting waste from other countries to Sweden. It also purposefully deflects attention away from the fact that it is a linear economy: the energy that waste produces is only used once. And it masks the oil and gas sector's heavy hand in increasing plastics production. It is also how the food industry 'bulks up' food by adding waste from other food products, such as COZ corn oil: waste is used as a food additive (van Tulleken, 2023: 226).

Yet, as we saw in the previous section, there are increasing calls for waste management technologies innovations and individual/community action to increase the collection of plastics waste in marine and land environments for recycling. These calls (strongly supported and often initiated by plastics companies detailed in the previous section) do not take into account the small proportion of plastics that can actually be recycled, and the environmental costs of recycling, including that recycling requires virgin fossil fuels and that it creates its own waste (see Chapter 4). Energy-from-waste facilities similarly create their own waste (which is much more toxic than the waste feeding the facilities, including furons and dioxins) and carry their own risks.

As Chapter 2 reflected, framing waste as an economic resource fits squarely within capitalist economies, which consider everything amenable to monetary value, calculation and profit. Whereas affluent societies still largely consider waste to be an inevitable excess of capitalism to be managed (hence the term 'waste management'), as primary resources dwindle waste itself is being 'folded back into' capitalist production (Lacy and Rutqvist 2016). It is leading, for instance, to urban mining whereby industries are assessing the profitability of extracting nickel, copper, iron/steel, aluminum, and other metals from urban infrastructure waste in a distinct form of urban mining, making use of what some industries refer to as 'hibernating stocks', since the amount of specific metals such as iron and copper in the built environment meets or 'exceeds the amount in known geological ores' (Johansson 2013: 1; Bergbäck and Lohm 1997; Spatari, et al. 2005). Pipes and cables constitute about a fourth of the weight of a city's infrastructure and in some cases contain as much or more sought-after metal as operating mines (Wallsten 2013). Resurrecting urks is a complex process, as cables and pipes laid down on top of each other over time through successive urban-space planning stages has created a vast, intricate, and difficult-to-access material system.

UNDERSTANDING WASTE AS SOCIAL INJUSTICE

A significant challenge to the dominant presentation of waste as either an individual post-consumption responsibility or as a profitable resource is understanding waste as social injustice. Framing waste as social injustice requires us to attend to macro- and micro- structures and practices that influence why our waste is managed in particular ways, and by whom (Hird, 2021; 2022; Hird and Predko 2024).

One of the salient ways in which waste is framed as social injustice is to reveal the relationship between waste and racism. Benjamin Chavis first used the term environmental racism in 1982 to refer to the polychlorinated biphenyl (PCB) waste that was disproportionately affecting a predominantly Black community situated near a landfill in Warren County, North Carolina (Fears and Dennis 2021). Chavis defined environmental racism as:

> racial discrimination in environmental policy making, the enforcement of regulations and laws, the deliberate targeting of communities of color for toxic waste facilities, the official sanctioning of the life-threatening presence of poisons and pollutants in our communities, and the history of excluding people of color from leadership of the ecology movements.
>
> (Chavis and Lee 1987: 3)

Understanding waste as social injustice means recognizing waste within the context of overarching critical issues involving historical and ongoing colonialism, poverty, and racialized and gendered inequality. Waste problems are far more likely to be found in or near racialized and lower-income communities because these communities are less likely to have access to, and influence within or over, larger government power structures that create and enforce waste regulations, nor municipal-level governments who choose where to site waste facilities such as landfills and incinerators. Racialized and lower-income communities, in any country, are far less likely to have members who are sufficiently income-secure to be able to volunteer their time to protest waste issues, as middle-class and wealthy communities do. And racialized and lower-income communities are, by definition, income-challenged, and are therefore more likely to agree to a waste facility near their community in exchange for money and/or other payments in kind such as the building of recreation centers, as well as promises of employment. And racialized and lower-income communities are far less likely to be able to afford the engineering and legal experts required to take industry and/or governments to court in order to force them to take legal and financial responsibility for contaminant remediation (for an example of how a largely white and relatively wealthy community was able to use its considerable financial and cultural capital to refuse the siting of a landfill near their community, see Forkert 2017).

One of the major aims of presenting waste as social injustice is to expose the relationship between the ever-increasing production and consumption demands of the world's privileged and the human health, environmental, labor, family and cultural consequences for the world's disadvantaged. As Chapter 2 details, clothing donations from developed countries overwhelm poorer countries, which has led some of these countries to ban second-hand clothing imports. This reality is also illustrated by the case of electronics. It is difficult to appreciate the planetary scale of the mining resource extraction required to furnish the raw materials – including rare earth materials such as ruthenium and indium – required for our electronic devices. As Chapter 1 details, colossal amounts of toxic waste are produced through extracting the required material. Then there is the energy required to manufacture the electronics, and the waste this manufacturing process produces. Add to this the waste produced in transporting these electronics around the globe, mainly in the form of carbon emissions from container shipping. Then using these electronics requires energy:

> Though individual usage may seem trivial, a simple internet request on Google equals the consumption of a 12 - watt light bulb for 2 hours; sending a 1 - megabyte attachment to ten correspondents requires the energy needed to move a car 500 meters. In an hour, around the world, 10 billion emails are sent, which corresponds to 50 gigawatt hours, or 4,000 Paris–New York round-trips by plane.
>
> (Jarrige and Le Roux 2020: 304)

The constant marketing of 'new and improved' electronics and their planned obsolescence results in increasing volumes of electronic waste. But while the appetite for electronics is increasing, our supplies cannot keep pace. Even if we were somehow able to recycle all of the metals and minerals contained in electronics, the current demand for electronics has already outpaced the supply that recycled electronic waste could produce. As electronics demand increases, the minimum gap to be filled by primary resource extraction simply increases as well. Documentaries such as *The E-waste Tragedy* (Dannoritzer 2014) and other exposés by Greenpeace (Weyler 2019) and human rights organizations detail the health impacts on poor Black/Indigenous/People of Color children, women and men as they disassemble these electronics; what David Naguib Pellow calls 'toxic colonialism' (2009). When discussions of waste are isolated from profound forms of inequality, then it is far easier to maintain the focus on individual responsibility.

REVIEW

Waste is primarily understood as the responsibility of individuals to reduce their own consumption and post-consumption waste, and as a problem that technological innovations in recycling and disposal will resolve. Individuals, families and households are held disproportionately responsible for the global waste problem despite post-consumption consumer waste accounting for the smallest proportion of waste generation. Conversely, the extractive and manufacturing industries produce far more waste (and often this waste is highly toxic) while taking little financial responsibility to either reduce or even remediate the waste they produce. Waste management companies and governments increasingly promote waste as an economic resource and a solution to ending energy dependence on fossil fuels. Burning waste produces energy that may be converted into electricity, and provides a way to solve a problem (waste) with a societal need and desire (energy consumption). Proponents do not focus on the waste that incineration produces, nor on the fact that the technology requires a constant supply of waste in order to function properly, and thus discourages waste reduction. Finally, grassroots organizations and people directly affected by pollution increasingly define waste as a social injustice problem that is only resolvable through increased producer responsibility and societal-level systemic changes to the current capitalist political-economic system.

SUGGESTED READING

Auyero, J. and Swistun, D.A. (2009). *Flammable: Environmental Suffering in an Argentine Shantytown*. Oxford University Press.

Bullard, R. and Wright, B. (2009). *Race, Place, and Environmental Justice After Hurricane Katrina*. Westview.

Mudu, P., Terracini, B. and Martuzzi, M. (Eds.) (2014). *Human Health in Areas with Industrial Contamination*. Copenhagen, Denmark: World Health Organization Regional Office for Europe.

Wiebe, S.M. (2016). *Everyday Exposure: Indigenous Mobilization and Environmental Justice in Canada's Chemical Valley*. University of British Columbia Press.

REFERENCES

Arthur, R. (2019). '*Don't* Buy *Coca-Cola* if You're Not Going to Help Us Recycle!' Coca-Cola Launches Recycling Campaign', *Beverage Daily*, 10 June. www.beveragedaily.com/Article/2019/06/10/Don-t-buy-Coca-Cola-if-you-re-not-going-to-help-us-recycle. Accessed 11 July 2024.

Bandoim, L. (2020). *Why Coca-Cola Refuses to Ban Plastic Bottle*, *Forbes Magazine*, 23 January. www.forbes.com/sites/lanabandoim/2020/01/23/why-coca-cola-refuses-to-ban-plastic-bottles/. Accessed 11 July 2024.

Bergbäck, B. and Lohm, U. (1997). Metals in Society, in D. Brune, D.V. Chapman, M.D. Gwynne, and J.M. Pacyna (eds.), *The Global Environment: Science, Technology and Management*. Scandinavian Science Publisher.

Berners-Lee, M. (2011) *How Bad are Bananas? The Carbon Footprint of Everything*. Greystone Books.

Bruner, R.G. (1990). The Plastics Industry and Marine Debris: Solutions Through Education, in Shonura, R.S. and Godfrey, L. (eds.), *Proceedings of the Second International Conference on Marine Debris, 2–7 April 1989*, pp. 1077–1089. https://swfsc-publications.fisheries.noaa.gov/publications/TM/SWFSC/NOAA-TM-NMFS-SWFSC-154.PDF. Accessed 11 July 2024.

Chavis, B. and Lee, C. (1987) *Toxic Waste and Race in the United States: A National Report on the Racial and Socio-Economic Characteristics of Communities with Hazardous Waste*. United Church of Christ's Commission on Racial Justice.

Chow, C.F., So, W.M.W., Cheung, T.Y., and Yeung, S.K.D. (2017). Plastic Waste Problem and Education for Plastic Waste Management, in Kong, S.C., Wong, T.L., Yand, M., Chow, C.F., and Tse, K.H. (eds.), *Emerging Practices in Scholarship of Learning and Teaching in a Digital Era*. Springer Singapore.

Crooks, H. (1993). *Giants of Garbage: The Rise of the Global Waste Industry and the Politics of Pollution Control*. James Lorimer & Company.

Dannoritzer, C. (2014). *The E-Waste Tragedy* [documentary]. RTVE Productions.

Deloitte. (2019). *Economic Study of the Canadian Plastic Industry, Market and Waste Task 5 – Summary Report to Environmental and Climate Change Canada*. Deloitte LLC.

Entman, R.M. (1993). Framing: Toward Clarification of a Fractured Paradigm, *Journal of Communication*, *43* (*4*): 51–58.

Fears, D. and Dennis, B. (2021). This Is Environmental Racism: How a Protest in North Carolina Farming Town Sparked a National Movement, *The Washington Post*, 6 April. www.washingtonpost.com/climate-environment/interactive/2021/environmental-justice-race/. Accessed 11 October 2023.

Forkert, P.G. (2017). *Fighting Dirty: How a Small Community Took on Big Trash*. Between the Lines.

Global Alliance for Incinerator Alternatives (GAIA) (2015). *Open Letter to Ocean Conservancy Regarding the Report 'Stemming the Tide'*. www.no-burn.org/wp-content/uploads/Open_Letter_Stemming_the_Tide_Report_2_Oct_15.pdf. Accessed 10 July 2024.

Government of UK. (2023). *UK Statistics on Waste*. www.gov.uk/government/statistics/uk-waste-data/uk-statistics-on-waste#total-waste-generation-and-final-treatment-of-all-waste. Accessed 11 October 2023.

Hinde, D. (2020). *The Swedish Recycling Revolution*. Government of Sweden. https://sweden.se/nature/the-swedish-recycling-revolution/. Accessed 14 January 2021.

Hird, M.J. (2021). *Canada's Waste Flows*. McGill-Queen's University Press.

Hird, M.J. (2022). *A Public Sociology of Waste*. Bristol University Press.

Hird, M.J. and Predko, H. (2024). *Extracting Reconciliation: Inhuman Wastes, Indigenous Lands, and Colonial Reckoning*. Routledge.

International Coastal Cleanup and Ocean Conservatory. (2015). *Stemming the Tide Report*. (Withdrawn and unavailable).

Jarrige, F. and Le Roux, T. (2020). *The Contamination of the Earth: A History of Pollutions in the Industrial Age*. The MIT Press.

Johansson, N. (2021). Intervention – 'Disaster Capitalism, COVID-19, and Single-Use Plastic', *Antipode Online*. https://antipodeonline.org/2021/01/26/covid-19-and-single-use-plastic/. Accessed 11 July 2024.

Johansson, N. (2013). *Why Don't We Mine the Landfills?* Linköping Electronic Press.

Judd, A. (2005). *In Defense of Garbage*. Praeger Publishers.

Kakadellis, S. and Rosetto, G. (2021). Achieving a Circular Bioeconomy for Plastics: Designing Plastics for Assembly and Disassembly is Essential to Closing the Resource Loop, *Science, 373 (6550)*: 49–50.

Kaza, S., Yao, L., Bhada-Tata, P., and Van Woerden, F. (2018). *What a Waste 2.0: A Global Snapshot of Solid Waste Management to 2050*. Urban Development Series. World Bank. https://doi.org/10.1596/978-1-4648-1329-0.

King, S. (2017). *My Week on a Plastic Beach Helping to Name and Shame Its Polluters*. Greenpeace Canada. www.greenpeace.org/canada/en/story/741/my-week-on-a-plastic-beach-helping-to-name-and-shame-its-polluters-2/. Accessed 10 July 2024.

Lacy, P., and Rutqvist, J. (2016). *Waste to Wealth: The Circular Economy Advantage*. Palgrave Macmillan.

Lepawsky, J. (2018). *Reassembling Rubbish: Worlding Electronic Waste*. MIT Press.

Liboiron, M. (2010) Recycling as a Crisis of Meaning, Etopia: Canadian Journal of Cultural Studies, 4: 1–9.

Liboiron, M. (2014). Solutions to Waste and the Problem of Scalar Mismatches, Discard Studies, 10 February. https://discardstudies.com/2014/02/10/solutions-to-waste-and-the-problem-of-scalar-mismatches/. Accessed 6 May 2020.

Liboiron, M. (2015). The Ocean Conservatory's Call for Mass Incineration in Asia: Disposability for Profit, Fantasies of Containment, & Colonialism, *Discard Studies*, 3 October. https://discardstudies.com/2015/10/03/the-ocean-conser vatorys-call-for-mass-incineration-in-asia-disposability-for-profit-fantasies-of-containment-colonialism/. Accessed 10 July 2024.

MacBride, S. (2012). *Recycling Reconsidered*. MIT Press.

Murphy, L. (2017). Sweden's Strange Problem: Not Enough Trash, Earth911, 3 January. https://earth911.com/business-policy/sweden-trash-problem/. Accessed 6 May 2021.

NPR. (2019). *The Litter Myth*. 5 September. www.npr.org/2019/09/04/757539617/the-litter-myth. Accessed 25 June 2024.

Ocean Blue Project. (n.d.). *Ocean Blue Project*. https://oceanblueproject.org/about-ocean-blue-project/. Accessed 10 July 2024.

Ocean Conservancy. (2022). *Stemming the Tide Statement of Accountability*. https://oceanconservancy.org/trash-free-seas/take-deep-dive/stemming-the-tide-statement-of-accountability/ Accessed 10 July 2024.

Owens, K. and Conlon, K. (2021). Mopping Up or Turning Off the Tap? Environmental Injustice and the Ethics of Plastic Pollution, *Frontiers in Marine Science*, 8: 1–8.

Pellow, D.N. (2009). Electronic Waste: The 'Clean Industry' Exports Its Trash, in C. Gossart (ed.), *Resisting Global Toxics: Transnational Movement for Environmental Justice*. MIT Press.

Rocher, L. (2020). Waste, a Matter of Energy: A Diachronic Analysis (1992–2017) of Waste-to-Energy Rationales, in Johansson, N. and Ek, R. (eds.), *Perspectives on Waste from the Social Sciences and Humanities: Opening the Bin*. Cambridge Scholars Press.

Rogers, H. (2006). *Gone Tomorrow: The Hidden Life of Garbage*. New Press; Signature Book Services.

Rugrod, P. (2023). Why Coastal Clean-ups Are Not Enough, *Greenpeace Thailand*, 16 September. www.greenpeace.org/southeastasia/story/62058/why-coastal-clean-ups-are-not-enough/. Accessed 10 July 2024.

Savini, F. (2019). The Economy That Runs on Waste: Accumulation in the Circular City, *Journal of Environmental Policy & Planning*. https://doi.org/10.1080/1523908X.2019.1670048. Accessed 11 July 2024.

Schnurr, R.E., Alboiu, V., Chaudhary, M., Corbett, R.A., Quanz, M.E., Sankar, K., Stain, H.S., Thavarajah, V., Xanthos, D., and Walker, T.R. (2018). Reducing

Marine Pollution from Single-Use Plastics (SUPs): A Review, *Marine Pollution Bulletin*, *137*: 157–172.

Spatari, S., Bertram, M., Gordon, R.B., Henderson, K., and Graedel, T.E. (2005). Twentieth Century Copper Stocks and Flows in North America: A Dynamic Analysis, *Ecological Economics*, *54*: 37–51.

State of California Department of Justice. (2024). Attorney General Bonta Sues ExxonMobil for Deceiving the Public on Recyclability of Plastic Products. 23 September. https://oag.ca.gov/news/press-releases/attorney-general-bonta-sues-exxonmobil-deceiving-public-recyclability-plastic#:~:text=SAN%20FRANCISCO%20—%20California%20Attorney%20General, the%20global%20plastics%20pollution%20crisis. Accessed 26 October 2024.

Supran, G., Rahmstorf, S., and Oreskes, N. (2023). Assessing ExxonMobil's Global Warming Projections, *Science*, *379* (*153*): 1–9.

Supran, G. and Oreskes, N. (2020). Reply to Comment on 'Assessing ExxonMobil's Climate Change Communications (1977–2014)', Supran and Oreskes (2017), *Environment Research Letters*, *15* (*11*): 118002.

US Environmental Protection Agency. (2022). *Containers and Packaging: Product-Specific Data*. www.epa.gov/facts-and-figures-about-materials-waste-and-recycling/containers-and-packaging-product-specific. Accessed 4 October 2023.

Urbancic, N., Harding-Rolls, G., Zallio, X.P.B., and Tangpuori, A.D. (2020). *Talking Trash: The Corporate Playbook of False Solutions to the Plastic Crisis*. Changing Markets Foundation. https://changingmarkets.org/report/talking-trash-the-corporate-playbook-of-false-solutions-to-the-plastic-crisis/. Accessed 11 July 2024.

van Tulleken, C. (2023). *Ultra-Processed People: Why We Can't Stop Eating Food That Isn't Food*. Penguin Random House.

Wallsten, B. (2013). *Underneath Norrköping: An Urban Mine of Hibernating Infrastructure*. PhD dissertation, Linköping University.

Walsh, A. (2023). Waves of Waste: The Harsh Truth About Ocean Plastic, *DW*, 13 November. www.dw.com/en/waves-of-waste-the-harsh-truth-about-ocean-plastic/a-67370326. Accessed 10 July 2024.

Watson, B. (2016). The Troubling Evolution of Corporate Greenwashing, *The Guardian*, 20 August. www.theguardian.com/sustainable-business/2016/aug/20/greenwashing-environmentalism-lies-companies. Accessed 26 October 2024.

Weyler, R. (2019). It's a Waste World, *Greenpeace*, 20 July. www.greenpeace.org/international/story/23747/its-a-waste-world/. Accessed 19 May 2021.

Wilson, D.C. and Velis, C.A. (2015). Waste Management: Still a Global Challenge in the 21st Century; an Evidence-Based Call for Action, *Waste Management & Research*, *33* (*12*): 1049–1051.

MANAGING WASTE

OVERVIEW

Chapter 4 details how waste is locally, nationally, and internationally managed. Municipalities, regions, countries, and supra-national organizations such as the European Union enact laws and devise policies to manage waste. Many countries and organizations have adopted the Waste Hierarchy and the Circular Economy to guide decision-makers in establishing legislation and policies that prioritize the environment and human health. On paper, reducing waste through prevention and minimization are the most favored ways of managing waste, while disposal, recycling, and waste-to-energy are the least favored. Waste management technologies, including engineered landfills, incinerators, waste-to-energy facilities, and recycling respond to the bottom levels of the Waste Hierarchy, and do the least to support the Circular Economy.

THE WASTE HIERARCHY

Throughout the world, waste is managed via a complex labyrinth of supra-national treaties and laws; national, regional, and local (municipal) regulations; and policies and practices. They are generally guided by the Waste Hierarchy (WH). The WH is an evaluation tool used to determine waste management options (see Figure 4.1).

DOI: 10.4324/9781003398424-4

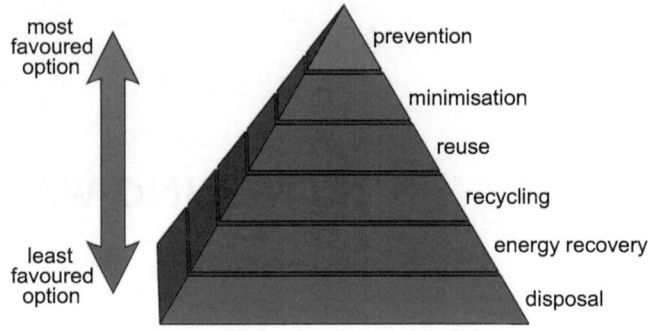

Figure 4.1 The Waste Hierarchy

Source: Wikipedia (Creative Commons, CC BY-SA 3.0)

It is scaled from most favorable to least favorable from an environmental perspective (Gharfalkar et al. 2015).

The WH refers to the end part of a product's life cycle, after the materials that the product requires have been extracted and the product has been manufactured, distributed, retailed, and consumed. From an environmental perspective, reducing waste through prevention (such as not buying new products) or minimization (such as consuming less) is the most favored option. Disposing of products (such as landfilling, open dumping, incineration) is the least favored option. In between are energy recovery (for instance, generating electricity from incineration), recycling (processing discarded products to manufacture new products), and reusing products.

At the supra-national level, the European Union's (EU) legally binding Waste Framework Directive (2008/98/EC) refers to prevention, reuse, recycling, recovery, and disposal. Individual countries such as Canada have also adopted the WH in strategic waste management planning (Government of Canada 2021). Other supra-national agreements manage waste with the intention of minimizing its negative effects on human health and the environment. For instance, various EU legislation targets particular waste streams, including: Animal By-Products Regulations; the Battery Directive; Cruise Ship Pollution; End-of-Life Vehicles Directive; Landfill Directive; Restrictions of Hazardous Substances Directive; and the Urban Waste Water Treatment Directive.

THE CIRCULAR ECONOMY

In tandem with the Waste Hierarchy, government organizations such as the European Union are turning to the Circular Economy (CE) with the aim of more effectively managing waste. As this section will show, waste is central to the CE. The CE is a model for extending the life cycle of products (European Parliament 2023). Whereas the linear economy is essentially an extract–produce–consume–waste process, the CE emphasizes minimizing products' environmental impacts, including waste. As Figure 4.2 illustrates, the CE is meant to incorporate every stage in a product's life cycle. It begins with product design, which entails designing products for maximum resource use efficiency, life-span, and reuse, as well as minimal (and ideally no) waste. More than eighty percent of a product's environmental impact is determined during the design phase. In the next stages, production and manufacturing, the goal here is to require the least amount of resource extraction as possible, and to produce products using the minimum of new material and the maximum of reused materials while creating the least amount of waste possible. Next, the CE focuses on the distribution of products, emphasizing local markets where possible, and the most efficient forms of global distribution. This means taking into account transportation (container ship, airplane, truck) carbon emissions and other environmental impacts. The next phase, consumption, focuses on reducing the amount of products that overconsuming people and societal sectors consume in order to ensure a much more equitable distribution of essential products globally. According to the World Bank, the richest ten percent of the world's population consumes almost sixty percent of all of the world's resources (Magdoff and Williams 2017). Then, building reuse and repair regulations, policies, and practices into consumption keeps products in use as long as possible, as opposed to the current proliferation of single- and short-use products that are soon discarded and replaced with new products, creating waste. This phase also targets planned obsolescence whereby products are intentionally designed to be quickly replaced, either by breaking easily or becoming unfashionable (see Chapter 2). The next steps, collection and recycling, focus on post–consumption product efficiency, which means diverting as many discarded products from disposal as possible and ensuring that these products are recycled into new products. Recycling is heavily

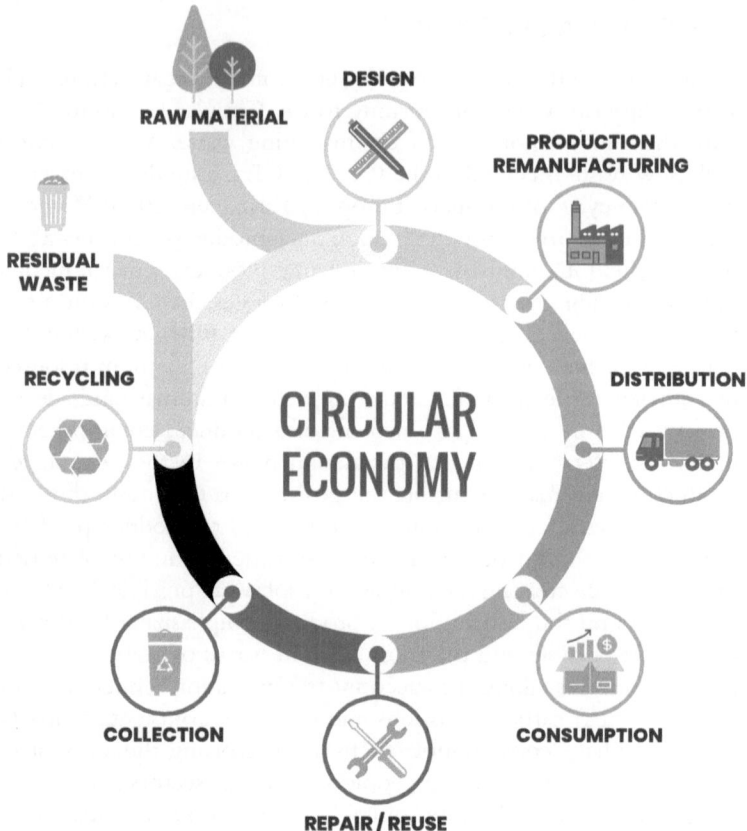

Figure 4.2 The Circular Economy

Source: Image by Freepik (www.freepik.com/free-vector/flat-design-circular-economy-nfographic_21095200.htm#query=circular%20economy&position=0&from_view=keyword&track=ais_hybrid&uuid=b5f3e922-b554-48b3-9173-6c9d98e1cd5e')

featured in CE models and yet, as the analysis below details, benefits the environment far less than is typically advertised. Finally, residual waste consists of all of the products and materials that cannot be kept within the circular loop. While the CE recognizes that a perfect loop is materially impossible (plastics is an example), the ultimate goal of the CE model is to minimize – as much as possible – the intake of raw materials and the output of residual waste.

In order to measure how circular our global economy is, the difference between new material and non-new material input into the system is required. This is a highly complex measurement to make, and must take into account biomass, fossil fuels (coal, oil, and gas), metals (copper, iron etc.), non-metallic minerals (such as gravel and sand), and land and water natural resources. The global economy consumes about 100 billion tons of materials per year (Circle Economy 2023). A small portion of this is circular. About 7.2 percent of all material inputs are secondary materials – this figure is down from 9.1 percent in previous years (Circle Economy 2023). Globally, our circular inputs are down not only because we are not cycling more, but significantly because we are extracting more virgin materials. We are, in other words, significantly increasing fossil fuels and minerals extraction. As Chapter 6 details, the transition to renewable (i.e. non-fossil fuels) energy is projected to lead to very high increases in material demand for lithium, copper, cobalt, graphite, nickel, and rare earth elements (Hund et al. 2020; Marscheider-Weidemann et al. 2021; Michaux 2021). In other words, we are making more products, and more products composed of more materials.

Fossil fuels (all of which are non-renewable) account for just over fourteen percent of inputs, while stock build-up (such as construction materials for buildings and infrastructure, as well as machinery) accounts for about thirty-eight percent of total material input (UNEP 2024). This very significant input is compounded by the fact that about twelve billion tons of the 43.6 billion tons of materials were wasted, as vehicles, machinery, appliances, and so on were trashed in 2018. Non-renewable inputs, such as metal ores and non-metallic minerals, are increasing, and are expected to increase by sixty percent from 2020 by 2060, from 100 billion to 160 billion tons. This use vastly exceeds the carrying capacity of the Earth as well as the 1.5 degree climate change limit (see Chapter 1). In all, this is creating 35.26 gigatons of waste and only 7.16 gigatons of circulated inputs back into the system.

Another portion of the material economy comes from renewable inputs, and in particular biomass (which may be carbon neutral or non-carbon neutral). Biomass accounts for about twenty-five percent of all material inputs (Circle Economy 2023).

There are numerous approaches to adopting a CE model. 'Slow strategies' focus on making products last as long as possible. There are material considerations here, from how long a given product is capable of high-performance functionality (an airplane, for instance), to how quickly products are designed to be replaced (fast fashion, for instance). 'Narrow strategies' focus on using less. Most of the global population is already using less, and is therefore creating less waste. 'Regenerate strategies' focus on replacing toxic materials with biomass resources. An example in farming is farmers being financially supported (through policy) to switch from synthetic to organic fertilizers, moving away from the use of growth hormones and pesticides (Circle Economy 2023). And 'cycling strategies' seek to maximize the amount and kinds of materials that remain in the economy, thereby minimizing the volume of virgin materials required. As the next section details, recycling products requires virgin materials, uses energy, and produces waste. Therefore, reusing and refurbishing products is environmentally superior to recycling. Cycling materials lays bare the fact that one hundred percent circularity is materially and technically impossible.

It is here that the Waste Hierarchy and Circular Economy converge. As Figure 4.3 illustrates, the highest levels of circularity are reached at the apex of the Waste Hierarchy: refusing (by not buying or otherwise consuming more than necessary to live), reducing, redesigning, repairing, and refurbishing all represent high levels of circularity, and provide the means for the world to live within the nine planetary boundaries and remain within the 1.5 degree Celsius global temperature increase. Regulations and policies that support energy recovery (such as energy-from-waste facilities), recycling, re-purposing, and re-manufacturing all result in low levels of circularity, and will not achieve sustainable living within planetary boundaries and restrict climate change.

The CE is considered at supra-national levels within organizations such as the European Union and the United Nations. At the national level, a number of countries are identifying the CE as a goal. The Japanese government, for example, aims to achieve 'full circularity' by 2025 through regenerative businesses that focus on product reduction, efficiency, and recycling (Kutty 2022). At the regional or city level, some cities are focusing on better designs for urban spaces that emphasize green and active transportation systems

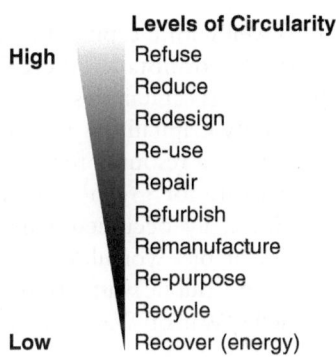

Figure 4.3 Levels of Circularity

Source: YouTube (www.youtube.com/user/carbontalks) (from Cropped screenshot from File:Towards a Circular City.webm (6:23) with UploadWizard – CC BY 3.0)

such as bicycling and walking, as well as public transport such as trams, light-rail trains, and buses (Ellen MacArthur Foundation 2019). Some cities are exploring the potential of what has been dubbed the Doughnut Economy. Similar to the CE, the Doughnut Economy focuses on meeting the needs of everyone within planetary boundaries (Raworth 2017; Goodwin 2021). As such, it draws from both the Planetary Boundaries Framework and the United Nations Sustainable Development Goals (see Chapter 6). The *Be-Circular* Brussels re-development plan, the *Paris Circular Economy Plan*, and the *Copenhagen Model* for circular bio-waste streams, and the Netherlands's *From Waste to Resource* are further examples (Savini 2019). A growing number of businesses are proclaiming adherence to CE principles, such as Patagonia, IKEA, Unilever, H&M, Adidas, Apple, and many more. Critics find that many companies promoting themselves CE adherents are actually referring to a limited part of their enterprises (see Stål and Corvellec 2018; Vonk 2018). And certainly, there are numerous microconsiderations of the CE that focus on individual and household practices of composting, recycling, sharing, down-sizing, and so on (see Chapter 6).

Like the Waste Hierarchy, the CE has its critics (for an overall synthesis see Corvellec et al. 2022). The first critique is that the CE is structured such that it fits well within capitalism (Savini 2019;Völker et al. 2020; Blum et al. 2020). For instance, the CE's focus on waste, Federico Savini (2019) argues, is capitalism's response to waste accumulation and resource scarcity. Capitalism's adaptation is to promote a strategy to 'valorize waste as a resource for city-regional growth' that in effect creates a structure for 'green' economic growth based on waste. Waste has traditionally occupied a negative position in society, as capitalism's inevitable worthless excess that must be removed from cities and the public's attention. This is illustrated by the European green-left's 'War on Waste' campaign in the 1980s (Cooper 2009).

As virgin material sources have become more scarce, and the costs of resource extraction have increased, the agenda has shifted to waste recovery as a means of economic development, and a source for further production. This is built into the CE model (see Figure 4.2). This explains, in part, why recycling has gained such popularity with governments and businesses. But the CE model, crucially, goes further than recycling in involving waste. Whereas recycling distinguishes between products that are waste and products (or their components) that are not actually waste, the CE defines waste itself as a resource separate from recycling. Focusing on waste as a resource keys into the regenerative and cycling strategies outlined above. In this way, individuals and households become 'prosumers' where 'consumption equals production through waste' (Savini 2019: 7). As such the CE acknowledges that household, construction, restaurant, grocery store, and other urban waste is increasing, and *valorizes* it as a commodity through which cities may continue to expand and profit. In other words, the CE extends capitalism's reach, and supports increasing waste production.

Another critique of the CE is that its definition is so varied that different stakeholders may choose the definition that best suits their ambitions. Kirchherr et al. (2018) inventoried over 100 definitions. Further, critics point out that the CE circle (or donut) implies that it can overcome the thermodynamic law of entropy. As Willi Haas (2019) points out, if we cycled all materials that can be cycled, it would only double the cycling rate (to nineteen percent). One third of all material is fixed in stocks (buildings, for instance). While we demolish buildings (liberating potential cyclable materials), we are building new infrastructure faster (thereby fixing more materials in stocks). As detailed above, another twenty-one percent of materials are non-cyclable fossil fuels. Moreover, some wastes do not have cycling options because they are contaminated or toxic (such as asbestos; see Johansson et al. 2020). As Cullen observes:

> Every loop around the circle creates dissipation and entropy, attributed to losses in quantity (physical material losses, by-products) and quality (mixing, downgrading). New materials and energy must be injected into any circular material loop, to overcome these dissipative losses.
>
> (2017: 483)

Finally, there is, as supra-national organizations such as the United Nations recognize, a profound difference in how people follow a CE approach based on income.

The per capita material footprint of high-income countries, the highest of all income groups, has remained relatively constant since 2000. Upper-middle-income countries have more than doubled their material footprint per capita approaching high-income levels, while their per capita impacts continue to be lower than high-income countries. Through global trade, high-income countries displace environmental impacts to all other income country groups. Per capita resource use and related environmental impacts in low-income countries has remained comparatively low and almost unchanged since 1995 (UNEP 2024: xiv).

This fact raises the critical issue of how we understand the problem, and therefore its solution (see Chapter 3). The CE is discussed by governments and industry in largely technocratic terms, as though technological innovations in energy recovery and recycling, coupled with consumer behavioral change, will effectively resolve both the global waste crisis as well as climate change, biodiversity loss and other planetary boundaries. In effect, it deflects attention away from, and in some cases silences, the socio-political and environmental injustice issues that the capitalist system created, maintains, and depends upon. For example, waste pickers – whose salvaging, repair, and reusing practices – are socially, economically, politically, and materially marginalized while at the same time serving a CE agenda. Consumers become 'prosumers' who are encouraged to consume more and more as waste is re-defined as a resource (Flynn and Hacking 2019). In sum, without taking seriously these critiques, the CE, like the Waste Hierarchy:

> risks turning into a hypothetico–normative (but self-serving) utopia that derails actual and well-intentioned efforts to reorganize production, consumption, and more generally material flows in ways that are more respectful of planetary boundaries and that work in favor of sustainability.
>
> (Corvellec et al. 2022: 429)

The case of plastics provides a sharp illustration of the vulnerability of both the Waste Hierarchy and the Circular Economy to work in service to the linear 'take-make-waste' industrial capitalist system. As Alice Mah notes, the plastics industry is 'very good at turning a crisis into an opportunity [and] the circular economy is a convenient way of doing so' (2021: 122). By explicitly preferencing reduction, refusal, and reuse, the CE and Waste Hierarchy threaten capitalist growth's 'business–as–usual' goal. Single- and short-use plastics bans and other regulations (see next section) similarly directly threaten the plastics industry's exponential profits. Research suggests that the plastics industry is finding ways to actually capitalize on the CE. Research shows that the plastics industry has long explicitly denied responsibility for plastics waste, deflecting responsibility on to consumers (Clapp 2012; O'Neill 2019; Allen et al. 2024). With the highly publicized discovery of plastics and microplastics in oceans around the world in the late 1990s, and governments turning their attention to the CE and WH as solutions, the petrochemical industry also focused on ways to utilize these models to their advantage. In 2018, on the heels of the European Commission's Strategy for Plastics in a Circular Economy, PlasticsEurope (a plastics lobby group) launched Plastics 2030, and in 2019, more than twenty-five plastics corporations (including Shell, ExxonMobil, and Dow) introduced the Alliance to End Plastic Waste. Much of this alliance was focused on diverting attention to sponsoring beach cleanups and recycling (Mah 2021; see Chapter 3), for which members of the public are held responsible (see Hird 2021), and by securing governments' acceptance of the industry's own targets, standards, and benchmarks for demonstrating their adherence to CE and WH goals. By 2019, the European Commission issued a bold press release declaring that in just three years, the Circular Economy Action Plan was complete. The report (*A Circular Economy for Plastics*), relied on consultations with industry experts (Mah 2021; Crippa et al. 2019). This report relies heavily on industry claims of advancing recycling technologies, including chemical recycling that – if even possible – is years away from being achieved, and even further away from mass implementation. In this way, the plastics industry ensures that a narrative of 'finding solutions' dominates at the same time that plastics production grows (Allen et al. 2024). And so the lower levels of the WH remain the focus, and the input of new (virgin) materials into the CE production-consumption system is obscured by heightened recycling promissory rhetoric.

CONVENTIONS, REGULATIONS, AND PROTOCOLS

Legislation and policies are needed to turn the Waste Hierarchy and Circular Economy models into action. Some legislation takes place at the supra-national level, while other legislation is specific to countries or regions.

The Basel Convention on the Control of Transboundary Movements of Hazardous Wastes and their Disposal is a supra-national agreement designed to eliminate the movement of hazardous waste between countries, and particularly from wealthy to poor countries. Adopted in March 1989 and ratified in May 1992, the convention specifically requires written consent from importers. A total of 189 countries, as well as the European Union, signed on to the treaty in 1989, but the United States and Haiti have yet to ratify it. In 2018, Norway proposed adding plastics waste to the convention's Annex VII in what has come to be known as the Basel Ban Amendment. To date, ninety-seven countries have ratified this amendment that bans the export of waste from a list of (including the European Union, countries within the OECD, and Liechtenstein) countries to (non-Annex VII) developing countries, and significantly includes recycling. Effectively, the amendment is designed to stop the export flow of waste from rich to poor countries under the guise of 'recycling' (see Chapter 5). Some countries, notably Australia, the United States, and Canada, objected to the Amendment.

One of the complexities that the Basel Ban Convention must contend with is that electronics are hard to define. Cars and airplanes have significant computer components. Fluorescent light bulbs contain circuit boards (and mercury). Yet, when discarded, they are categorized as electronic waste. And because definitions of electronics vary, it is also difficult to distinguish between electronics destined for disposal from electronics destined for reuse, refurbishment, and/or recycling (Lepawsky 2018: 10). The Basel Ban Convention distinguishes between Annex VII countries (country members of the OECD, EU, and Liechtenstein) and non-Annex VII countries (any country that is not an Annex VII country). While the Basel Ban restricts flows of e-waste from Annex VII to non-Annex VII countries, most e-waste is actually exported within Annex VII countries (for instance, from the United States to Mexico and Canada). Non-Annex VII countries mainly trade to other Annex VII countries. This is due to the fact that electronics materials are valuable, and so most electronics are reused, refurbished, and recycled rather than disposed of after being discarded. The top ten electronics waste importers include France, Sweden, the United Kingdom, Belgium, and Germany. Africa is a net e-waste exporter. Most electronics waste trade in Asia is from the Republic of Korea to Japan. The popular narrative that poor countries are overwhelmed by e-waste imports obscures the fact that most e-waste is produced in the making of electronics, rather than after they are discarded (see Chapter 5).

Besides the Basel Ban Convention and its Amendment, other supra-national initiatives are designed to manage particular kinds of waste. Other international laws include the Stockholm Convention on Persistent Organic Pollutants (2001), the Convention on the Prior Informed Consent Procedure for Certain Hazardous Chemicals and Pesticides in International Trade (1998), Convention on the Transboundary Effects of Industrial Accidents (1992), and the Convention on Civil Liability for Damage Caused During Carriage of Dangerous Goods by Road, Rail, and Inland Navigation Vessels (1989), among others.

In 2018, Canada led the G7 in developing the Open Plastics Charter, establishing a framework for sustainable plastics use and eliminating plastics litter on land and at sea. Globally, twenty-eight governments and over seventy companies and organizations have endorsed the Charter. In March 2022, the UN Environment Assembly adopted a resolution to combat plastics pollution with a global and legally binding plastics treaty by 2024 that will take into consideration the whole plastics life cycle with internationally binding targets, called the Internationally Legally Binding Instrument on Plastic Pollution, Including the Marine Environment (Bergmann et al. 2022; UNEA UNEP 2022; see Chapter 6). The Gothenburg Protocol (The United Nations Convention on Long-Range Transboundary Air Pollution, or CLRTP) refers to ammonia emissions (UNECE, 1999). The EU Nitrates Directive (EEC, 1991) refers to nitrate and phosphorous water pollution. And the Kyoto Protocol regulates methane and nitrous oxide water pollution as part of the UN Framework Convention on climate change (United Nations 1997).

As well, individual countries have their own legislation. The United Kingdom, for instance, enacted the Environment Act 1995, the Landfill in the UK, Landfill Tax Regulations, Landfills in the UK, Control of Pollution Act, Waste Management Licensing Regulations, and other legislation that impacts waste prevention and management. And within individual countries, provinces, states, departments, and other regions enact laws, regulations and policies that manage waste. For instance, California's Proposition 65, The Safe Drinking Water and Toxic Enforcement Act of 1986, prohibits the discharge of toxic substances into drinking water sources. Canada's federal government produces the National Pollutant Release Inventory, which is a publicly accessible database containing information about over 300 pollutant releases, transfers, and disposals that over 7,000 facilities across Canada are required to report (Berthiaume 2023).

Individual countries may classify waste in ways that correspond with, or do not correspond with, other countries. For instance, France has recently re-classified some intermediate-level radioactive waste as low-level waste, which means that, in France, this radioactive waste may be mixed with MSW and disposed of in landfills or incinerators that are sited close to communities and have been engineered for household waste (Garcier 2014). And in some instances, international and national laws coincide with, or contradict, each other. For example, discharge limit values (DLVs) are amounts of pollutant concentrations in effluent or discharged pollution load that must not be exceeded within a certain time period. DLVs are typically legally binding (Preisner, Neverova-Dziopak, and Kowalewski 2020). The standardization of DLVs and wastewater quality is complex because of numerous factors, including diverse legislation (which itself leads to different licensing systems for wastewater treatment facilities, different prohibitions and restrictions, legal liabilities, penalties, and so on) for regulating threshold levels for contaminants in effluent discharge and methodologies for measuring effluent discharge and its effects. Numerous waterways are shared by several countries, and may or may not be subject to international regulations as well as the regional and national laws of the countries sharing the water system (for a discussion of what happens when international laws clash with national laws, see Howarth and Marino 2006).

In sum, the treaties, agreements, laws, regulations, and policies that govern how all types of waste (from MSW to radioactive) are managed are made at the global level by institutions such as the United Nations. Supra-national entities also manage waste, such as the East African Community (comprising Burundi, Kenya, Rwanda, South Sudan, Tanzania, and Uganda) that has banned the import of used clothing since 2019. Within countries, waste laws, regulations, and policies are developed and implemented at national, regional, and local levels.

WASTE MANAGEMENT TECHNOLOGY

Along with numerous and varied supra-national, national, regional, and municipal regulations and policies that govern how waste is managed, a number of technologies materially manage waste. These technologies variously minimize waste volumes (for instance, incineration) or remove waste from within communities (for instance, landfills).

DUMPS AND LANDFILLS

Open dumping occurs when people discard municipal solid waste in open-air, non-engineered conditions. According to UNEP (2024), about forty percent of the world's waste is openly dumped, which amounts to almost one billion tons annually. Some two billion people have no waste collection services and at least three billion people cannot access waste disposal facilities (Wilson and Velis 2015). Open dumping most frequently occurs in lower income areas. For instance, open dumping is the norm in Canada's Arctic (Hird 2021). It is the cheapest way to deal with waste, but comes with potentially high adverse human health effects and environmental contamination. In some regions, communities are built on waste dumps (Amegah and Jaakkola 2016; Yang and Furedy 1993; Mothiba et al. 2017;

Parizeau 2006). Mega slums around the globe are increasing in number and size: Bantar Gebang in Indonesia is a twenty-eight to forty
million ton open dump that accepts some 230,000 tons of waste per
year (ISWA 2016); Jam Chakro in Pakistan covers some 500 acres
and has around five million people living in its vicinity. A number
of studies (for example Dias 2016; Dias and Fernandez 2013) exam-
ine the highly dangerous job of waste picking in what Mike Davis
describes as our 'planet of slums', in his book of the same title (2007).

In contrast to dumps, landfills are engineered. Landfills are popular
in countries like the United States, Canada, China, and Russia
because of the large amount of available land. Much smaller countries
such as France, the United Kingdom, and Portugal also continue to
somewhat rely on landfills, although incineration is more favored.

The aim of the modern engineered landfill is to keep its contents
in place, as static as possible, and for as long as possible. Engineered
landfills are composed of three primary subsystems of barrier. Aerobic bacteria metabolize during the early life of a landfill, which
produces material that is highly acidic and toxic to surface water.
Anaerobic bacteria do the bulk of the metabolizing work deeper in
the landfill's strata, producing leachate. And this leachate may travel
vertically and horizontally within landfills and may continue to
travel when it leaks out of the landfill. That is, leachate may percolate
into soil and groundwater, where it moves into and through plants,
trees, animals, fungi, insects, and the atmosphere. It is this leachate
that is also responsible for the methane that emanates from landfills.
In 2010, landfills around the world produced nearly 882 million
tons of emissions, which amounts to about eleven percent of all
human-generated methane (Gies 2016). Landfill operations and
cover require considerable attention to siting, design, construction,
operations, and post-closure care (from decades to centuries).

Landfill regulations do not always address issues such as 'contaminants of emerging concern' (Celik et al. 2009; LaPensee et al. 2009;
Rowe 2012; Takai et al. 2000). Contaminants of emerging concern include chemicals such as bisphenol A, which has been used
in many plastic products and is believed to mimic human estrogen
at low concentrations (LaPensee et al. 2009; Takai et al. 2000), and
polybrominated diphenyl ether, which is an additive flame retar-
dant in plastics, foams, and fabrics that may cause liver, thyroid, and
neurodevelopmental toxicity – as well as new materials such as

nanoparticles, which were not part of the waste stream at the time many landfill regulations were developed (such as US Subtitle D) (Islam and Rowe 2009; Rowe 2012; LaPensee et al. 2009; Takai et al. 2000). In addition, calculating the environmental costs must include the climate change implications of using fossil fuels to transport waste to landfills (see Chapter 3).

A significant challenge facing landfill engineering is that waste increasingly contains various amounts and kinds of seven million or so known chemicals (as well as about a thousand new chemicals that enter into use each year), along with a full spectrum of organic matter, which includes the 14,000 food additives and the contaminants found in our food scraps, coal fly ash (of which over 50 per cent ends up landfilled; see Chertow 2009), plastics (more than 390 million tons of plastics are consumed worldwide each year, most of which still end up landfilled; see PlasticsEurope 2018), and food waste (over ninety-seven per cent of which is landfilled in the United States; see Levis et al. 2010). The liquid material, called leachate, into which organic landfill dissolves frequently consists of a heterogeneous mix of heavy metals, endocrine-disrupting chemicals, phthalates, herbicides, pesticides, and various gases, including methane, carbon dioxide, carbon monoxide, hydrogen, oxygen, nitrogen, and hydrogen sulphide (Hird 2021, 2022).

To provide one example of the diverse material that may end up in landfills, XL Foods, Canada's largest food processor, processes over forty per cent of the country's cattle and accounts for 30 per cent of the beef on store shelves. In 2012, approximately 5.5 million kilograms of beef presumed to be contaminated with E. coli was recalled, equivalent to 12,000 cattle. Of that, 500,000 kilograms were landfilled. XL Foods were also required to do a pilot test to ensure their corrective measures after the recall were effective. This test required the slaughter of 5,000 cattle, the carcasses of which were also landfilled after being tested for contamination, regardless of whether they had actually been contaminated (Lougheed et al. 2016; Lougheed 2017).

INCINERATION AND WASTE-TO-ENERGY FACILITIES

While, globally, open dumping and landfilling waste remain the most used forms of waste management, many countries employ incineration. Traditional incinerators burn waste at very high temperatures.

Incinerator gases entering the atmosphere via incinerator stacks may produce particulate pollution associated with cardiovascular and cerebrovascular mortality. Many of these emissions occur during day-to-day facility operations as well as the 'very high releases of dioxins that arise during start up and shutdown of incinerators' (Thompson and Anthony 2008: 2). Moreover, all incinerators produce waste ash that is potentially more hazardous than the waste that feeds the incinerator (Rowe 2012). This waste ash is landfilled, which means concentrated levels of heavy metals are buried in the ground amid other landfilled materials as well as other constituents (such as calcium) that accelerate leachate collection system clogging, increasing the risk of leachate leakage. When ash hydrates within landfills, it generates substantial heat that could compromise liner systems that are typically not designed to accommodate these temperatures.

In order to function properly, incinerators require a constant input of waste. For this reason, countries such as Sweden and Denmark import waste from other countries (increasing carbon emissions via the transportation of this waste), and some municipalities encourage their citizens to recycle less waste in order to have sufficient waste for incineration. As one incinerator operator said, responding to China's plastics recycling import ban, 'if we do the plastic ban, we would have to look for other more distant waste' (in Bahers 2021: 10). In addition, waste-from-energy technologies are linear, not circular: once the energy (electricity and/or heating) is used, it is gone. Some waste-to-energy facilities are intended to attract attention, and encourage further consumption. For instance, the Amager Bakke waste-to-energy facility in Copenhagen, Denmark, burns municipal, industrial, construction, and institutional waste collected from the region as well as waste imports from countries such as the United Kingdom (Murray 2019). It is also purposefully designed to draw crowds of visitors to re-conceptualize waste as a benign amusement park. The waste facility is also a ski slope, a running path, an 80-meter-high artificial climbing wall, a café, and a tree grove. As such, a key part of energy-from-waste technologies is that they encourage further consumption. This is also the case with recycling. It licenses people to consume more because companies and governments present recycling as an effective way to minimize waste to the public.

RECYCLING

Open dumping, landfills, and incinerators all appear at the bottom of the WH because they manage waste exclusively through disposal. Waste-to-energy facilities appear further up the WH because they generate energy (as well as generate waste), and as such manage waste with more than simply disposal. Recycling post-consumption waste also appears further up the WH than disposal. Of all the materials that are diverted from landfills, food waste is unquestionably the most environmentally beneficial to recycle because food waste can be transformed into soil that can then be used to grow more food. The more local the composting, the better for the environment, since transporting food waste to composting facilities is usually done with trucks that use non-renewable fossil fuels and are thus contributing to climate change. Diverting food waste from landfills is particularly beneficial because food waste is especially attractive to various kinds of aerobic bacteria that transform this waste into leachate that, combined with the various hazardous materials outlined above, and leaked into the environment, can cause serious human health and environmental harm.

Apart from recycling food waste, recycling other materials is not necessarily beneficial to the environment. First of all, recycling is a profit-driven enterprise. The materials that individuals and households separate into recycling bins are only actually recycled if the recycling company makes a profit. Due to constantly varying markets, waste intended for recycling is often disposed of when recycling costs outweigh the profit derived from the recycled materials. Some recycling companies have the capacity to stockpile some materials while they wait for a more favorable market, but most often recycling companies simply move the materials into the disposal (landfill or incineration and then landfill for the incinerator ash) stream. Moreover, few regulations require recycling companies to declare what proportion of the materials they collect are actually recycled, and local governments are typically not legally required to monitor recycling companies – that is, once recycling companies collect materials, municipal governments are not required to know where the recycling companies move the materials, including whether the materials are taken to a recycling facility or to a landfill or incinerator. Without an ongoing market for many recycled materials, the 'consequence is that materials thought by the public to be headed for recycling end up in landfills' (Rowe 2012: 6). For this reason, PlasticsEurope (2018) estimates that the vast bulk of plastics ends up landfilled.

Second, the physical process of recycling materials has a negative environmental impact that must be measured against the environmental benefits of recycling. Recycling materials consumes a great deal of energy and mainly entails using non-renewable fossil fuels that pollute the soil and atmosphere and contribute to global warming (Greenpeace Canada 2020; MacBride 2012). Mechanical plastics-to-plastics recycling involves 'sorting, washing, shredding/ grinding, melting and pelletizing plastics waste into secondary raw material' (Greenpeace Canada 2020: 13). Not only is this most popular form of plastics recycling unable to repair recycled plastic degradation, leading to a reduction in recycled plastics quality and therefore reusability, but this has led to the mixing of virgin resin, which is derived from oil and gas. Thus, mechanical plastics recycling leads to an *increase* in oil and gas extraction.

For example, a study by Jorge Vendries et al. (2020) conducted a meta-analysis of the results of available research conducted on the life cycle of various packaging and food service ware materials. The researchers were interested in determining whether particular material attributes such as recycled content and recyclability produce lower net environmental impacts across the full life cycle of the packaging and food service ware. Analyzing the results of seventy-one studies that included over 5,000 comparisons of thirteen impact categories, the findings were illustrative of the complexities of recycling. The impact categories included: human toxicity, global warming, fossil energy, ecotoxicity, eutrophication, smog, acidification, PM formation, ozone depletion, mineral depletion, water consumption, land occupation, and ionizing radiation. The best-case scenario occurred with the same material that contained recycled content. In this case, just twenty percent showed significantly lower impacts across all categories, while seventy-four percent demonstrated marginally lower impacts (Vendries et al. 2020: 5359). That is, when only one material is chosen, it has only a twenty percent lower environmental impact when more of that material comes from recycled content than from non-recycled (that is, new) content.

The meta-analysis significantly revealed that the material matters more than recycled content in terms of environmental impacts: '[I]n fact, of 534 comparisons of packaging made from different materials and with different levels of recycled content, the packaging with higher recycled content had significantly higher impacts in 60% of the comparisons and significantly lower impacts in 21%' (Vendries et al. 2020: 5359). The results for recyclable packaging content using different material, biobased packaging using either the same material or different materials, compostable versus non-compostable packaging, and composting packaging versus other end-of-life options demonstrate a net significant *cost to* (that is, *negative* impact on) the environment. According to a recent life-cycle analysis study of the environmental footprint of polystyrene, researchers Brooke Marten and Andrea Hicks (2018) note that just the re-expanding and shaping stages of polystyrene recycling takes 30 per cent of the energy used in the life cycle of this material.

Third, transporting waste to reach recycling processing facilities significantly mitigates claims that recycling is always the environmentally better option. As William Baarschers notes, 'after the diesel fuel is burned, the carbon dioxide and nitric oxides produced, we have waste tires, waste batteries, and waste trucks at the end of the line' (1996: 190). In other words, recycling that relies on mid- and long-haul transportation contributes to the problem of global warming. According to Naya Olmer et al.'s (2017) analysis of container fleet carbon dioxide emissions, container ships alone contribute about twenty-three percent of human-produced carbon emissions (more than oil tankers, which account for thirteen percent). In 2015, international container ships produced approximately 31,419 million tons of carbon dioxide.

Fourth, recycling may release hazardous wastes into the environment through by-product emissions and/or require the use of toxic materials. Recycling paper, for instance, requires the significant use of toxic chemicals to remove dye, and then re-dye the paper for a second use. And paper cannot be recycled indefinitely: each recycling degrades the product until it is no longer recyclable (Baarschers 1996).

Fifth, products made from recycled materials are generally only suitable for one or two subsequent uses and usually only in lesser quality products. This is what is known as the 'low-value limit'. As Jodie Morgan observes: '[P]lastic in general is a fairly low-value product… we spend a lot of money collecting that product, sorting that product, processing that product all so it can go back into… a relatively low-value end product' (in Chung 2019). While recycling rhetoric and advertising promotes the assumption that products are recycled indefinitely, the reality is that, of the products that actually remain in the recycling stream, many are only recycled once or twice before being landfilled or incinerated.

Sixth, a lot of the plastics waste that we put in the recycling stream *cannot* materially be recycled. Many types of non-durable plastics, such as those commonly used in food packaging, may be put in recycling bins by consumers, but they are put back into the disposal stream when they are transported to a recycling sorting center (National Geographic 2018).

Sixth, recycling has not led to a decrease in resource extraction and manufacturing – which was the entire intent of the recycling promise. The case of electronics provides a stark example. Even if we were able to recycle all of the metals and minerals contained in electronics such as laptops, smart phones, and cars, the current demand for electronics has already outpaced the supply that recycled electronic waste could produce. As electronics demand increases, the minimum gap to be filled by primary resource extraction simply increases as well. Moreover, evidence suggests that people actually produce more waste when they have access to recycling (Harris 2015; Hird 2021).

WASTE MANAGEMENT RESPONSIBILITY

Some of the laws, regulations, and policies cited at the beginning of this chapter refer to pre-consumer waste. That is, waste produced by resource extraction, manufacturing, distribution, and retail is regulated regionally, nationally, and globally. However, as Chapter 1 describes, the waste that people are most familiar with is the waste we produce as consumers: post-consumption waste.

Beyond the material and economic limitations outlined above, the biggest problem with recycling is that it draws attention away from holding companies that manufacture single- and short-use products and packaging. In other words, recycling deflects attention away from extended producer responsibility regulations and policies, which require product and packaging producers to take financial and material responsibility for the waste they create. As such, recycling works very well for the extractive and manufacturing industries and for retailers, and has negative impacts on consumers, citizens, and the environment. Recycling reveals a significant imbalance of power between extractive, manufacturing, and retail industries and the public, who are shouldering the financial burden of industry's waste production.

There is a clear causal association between consumption and waste production. And there is, as Kathryn Wheeler and Miriam Glucksmann point out, a strong moral economy that operates such that the 'responsible "citizen-consumer" is motivated to act because of his or her commitment to moral/political projects rather than in line with his or her selfish desires' (2015: 143–144). Many countries have adopted recycling as a central part of their moral economy and emblematic of the 'good environmental citizen': for instance, studies show that countries such as the United Kingdom and Sweden have adopted this discourse wholesale (Wheeler and Glucksmann 2015; Skill 2008). Indeed, Skill (2008) demonstrates that around the globe, 'recycling is the most common action that households regularly performed' (in Wheeler and Glucksmann 2015: 153). In Sweden, for instance, Swedish-raised respondents distinguished themselves from immigrants as the 'irresponsible other', based on the perceived lack of proper recycling performed by immigrants (Sayer 2000). Around the globe, then, ordinary people are shouldering the physical, financial, and moral burden of a system that monetarily favors companies and has negative environmental and human health consequences, deflects attention away from the much greater volumes of pre-consumption waste produced by resource extraction and manufacturing industries, and presents as an effective solution to our global waste crisis without any need to change the current political-economic system that both enables and rewards waste production.

REVIEW

Many countries and environmental organizations have adopted the Waste Hierarchy and the Circular Economy to guide their waste management priorities. Focused on the human health and environmental impacts of waste, the WH prioritizes waste reduction, followed by reuse/refurbishment, waste-to-energy, and finally recycling and disposal. Disposal and recycling remain the most common ways of managing waste worldwide. In order to manage waste more effectively, countries need to prioritize reduction over disposal and recycling.

SUGGESTED READING

Greer, J. and Bruno, K. (1997). *Greenwash: The Reality Behind Corporate Environmentalism*. Rowman and Littlefield.

Hird, M.J. (2022). *A Public Sociology of Waste*. Bristol University Press.

MacBride, S. (2012). *Recycling Reconsidered*. MIT Press.

Pierre-Louis, K. (2012). *Green Washed: Why We Can't Buy Our Way to a Green Planet*. Ig Publishing.

REFERENCES

Allen, D., Linsley, C., Spoelman, N. and Johl, A. (2024). *The Fraud of Plastic Recycling: How Big Oil and the Plastics Industry Deceived the Public for Decades and Caused the Plastic Waste Crisis*. Center for Climate Integrity.

Amegah, A.K. and Jaakkola, J.J. (2016) Household Air Pollution and the Sustainable Development Goals, *Bull World Health Organ*, *94* (*3*): 215–221.

Baarschers, W.H. (1996). *Eco-Facts and Eco-Fiction: Understanding the Environmental Debate*. Routledge.

Bahers, J.B. (2021). 'en quoi les politiques locales d'economie circulatire no sont pas des resistances aux regimes metaboliques dominants: le cas des metabolismes urbains de nantes-nazaire et goteborg', Paper presented at the Congres ABSP-CoSPoF conference, 8 April. Bruxelles, France. Bergmann, M., Almroth, B.C., Brander, S.M., Dey, T., Green, D.S., Gundogdu, S., Krieger, A., Wagner, M., and Walker, T.R. (2022). A Global Plastic Treaty Must Cap Production, *Science*, *376* (*6592*): 469–470.

Berthiaume, A. (2023). New Sustainability Perspectives on Pollutant Releases from Canada's Nuclear Sector, *Environmental Science and Technology*, *57* (*35*), 12958–12968.

Blum, N.U., Haupt, M., and Bening, C.R. (2020). Why 'Circular' Doesn't Always Mean 'Sustainable', *Resources, Conservation and Recycling*, *162*: 105042.

Celik, B., Rowe, R.K., and Unlü, K. (2009). Effect of Vadose Zone on the Steady- State Leakage Rates from Landfill Barrier Systems, *Waste Management*, *29* (*1*): 103–109.

Chertow, M. (2009) The Ecology of Recycling, *UN Chronicle*, *46* (*3–4*): 56–60.

Chung, E. (2019) Most Styrofoam Isn't Recycled: Here's How 3 Startups Aim to Fix That, *CBC News*, 25 March. www.cbc.ca/news/technology/styrofoam-chemical-recycling-polystyrene-1.5067879. Accessed 20 April 2021.

Circle Economy. (2023). *The Circularity Gap Report 2023*. Circle Economy and Deloitte.

Clapp, J. (2012). The Rising Tide Against Plastic Waste: Unpacking Industry Attempts to Influence the Debate, in Foote, S. and Mazzolini, E. (eds.), *Histories of the Dustheap: Waste, Material Cultures, Social Justice*. MIT Press.

Cooper, T. (2009). War on Waste? The Politics of Waste and Recycling in Post-War Britain, 1950–1975, *Capitalism Nature Socialism, 20* (4): 53–72.

Corvellec, H., Stowell, A., and Johansson, N. (2022). Critiques of the Circular Economy, *Journal of Industrial Ecology, 26*: 421–432.

Crippa, M., De Wilde, B., Koopmans, R., Leyssens, J., Linder, M., Muncke, J., Ritschkoff, A-C., Van Doorsselaer, K., Velis, C., and Wagner, M. (2019). *A Circular Economy for Plastics: Insights from Research and Innovation to Inform Policy and Funding Decisions*, M. De Smet and M. Linder (eds.). European Commission.

Cullen, J.M. (2017). Circular Economy: Theoretical Benchmark or Perpetual Motion Machine? *Journal of Industrial Ecology, 21* (3): 483–486.

Davis, M. (2007) *Planet of Slums*. Verso Books.

Dias, S.M. (2016). Waste Pickers and Cities, *Environment and Urbanization, 28* (2): 375–390.

Dias, S. and Fernandez, L. (2013). Wastepickers: A Gendered Perspective, in *Powerful Synergies: Gender Equality, Economic Development and Environmental Sustainability*. United Nations Development Programme.

EEC. (1991). *Directive 91/676/EEC of 12 December 1991 Concerning the Protection of Waters Against Pollution Caused by Nitrates from Agricultural Sources*. Off. J. Eur. Comm. L375, 1–8. EEC.

Ellen MacArthur Foundation. (2019). *City Governments and Their Role in Enabling a Circular Economy*. https://emf.thirdlight.com/link/lg3ap956qxbi-66omej/@/#id=0. Accessed 8 July 1024.

European Parliament. (2023). *Circular Economy: Definition, Importance and Benefits*. www.europarl.europa.eu/topics/en/article/20151201STO05603/circular-economy-definition-importance-and-benefits. Accessed 8 July 2024.

European Union. (2008). *Waste Framework Directive*. https://eur-lex.europa.eu/legal-content/EN/TXT/?uri=celex%3A32008L0098. Accessed 25 June 2024.

Flynn, A. and Hacking, N. (2019). Setting Standards for a Circular Economy: A Challenge Too Far for Neoliberal Environmental Governance? *Journal of Cleaner Production, 212*: 1256–1267.

Garcier, R. (2014). Disperse, Confine or Recycle? A Geo-logical Approach to the Management and Spatial Circulations of Low-Level Radioactive Waste in France, *L'Espace Géographique, 43* (3): 265–283.

Gharfalkar, M., Court, R., Campbell, C., Ali, Z., and Hillier, G. (2015) Analysis of Waste Hierarchy in the European Waste Directive 2008/ 98/ EC, *Waste Management, 39*: 305–313.

Gies, E. (2016) Landfills Have a Huge Greenhouse Gas Problem: Here's What We Can Do About It, *Ensia*, 25 October. https://ensia.com/features/methane-landfills/. Accessed 21 April 2021.

Goodwin, K. (2021). *Designing the Doughnut: A Story of Five Cities*. Doughnut Economics Action Lab (DEAL). https://doughnuteconomics.org/stories/designing-the-doughnut-a-story-of-five-cities. Accessed 8 July 2024.

Government of Canada. (2021). Reducing Municipal Solid Waste. www.canada.ca/en/environment-climate-change/services/managing-reducing-waste/municipal-solid/reducing.html. Accessed July 30 2023.

Greenpeace Canada. (2020). Plastic Recycling: That's Not a Thing, www.greenpeace.org/static/planet4-canadastateless/1d30117a-greenpeacereport_plasticrecyclingthatsnotathing.pdf. Accessed 5 January 2021.

Haas, W. (2019). From Throwaway Society to Circular Economy: Solution or Comforting Illusion? *EuropeNow*, 7 May. www.europenowjournal.org/2019/05/06/from-throwaway-society-to-circular-economy-solution-or-comforting-illusion/. Accessed 27 October 2024.

Harris, S. (2015). Canadians Piling Up More Garbage Than Ever Before as Disposables Rule, *CBC News*. www.cbc.ca/news/business/canadians-piling-up-more-garbage-than-ever-before-as-disposables-rule-1.3248949. Accessed 27 October 2024.

Hird, M.J. (2022). *A Public Sociology of Waste*. Bristol University Press.

Hird, M.J. (2021). *Canada's Waste Flows*. McGill-Queen's University Press.

Howarth, R.W. and Marino, R. (2006) Nitrogen as the Limiting Nutrient for Eutrophication in Coastal Marine Ecosystems: Evolving Views Over Three Decades, *Limnol Oceanography, 51*: 364–376.

Hund, K., La Porta, D., Fabregas, T.P., Laing, T., and Drexhage, J. (2020). *Minerals for Climate Action: The Mineral Intensity of the Clean Energy Transition – Report*. The World Bank Group. http://pubdocs.worldbank.org/en/9617115888-75536384/Minerals-for-Climate-Action-The-Mineral-Intensity-of-the-Clean-Energy-Transition.pdf. Accessed 8 July 2024.

Islam, M.Z. and Rowe, R.K. (2009) Permeation of BTEX Through Unaged and Aged HDPE Geomembranes, *Journal of Geotechnical and Geoenvironmental Engineering, 135 (8)*: 1130–1140.

ISWA. (2016). A Roadmap for Closing Waste Dumpsites, the World's Most Polluted Places. www.resource-recovery.net/sites/default/files/iswa_dumpsites-roadmap_report.pdf. Accessed 19 May 2021.

Johansson, N., Velis, C., and Corvellec, H. (2020). Towards Clean Material Cycles: Is There a Policy Conflict Between Circular Economy and Non-Toxic Environment?, *Waste Management and Research, 38 (7)*: 705–707.

Kirchherr, J., Piscicelli, L., Bour, R., Kostense-Smit, E., Muller, J., Huibrechtse-Truijens, A., and Hekkert, M. (2018). Barriers to the Circular Economy: Evidence from the European Union (EU), *Ecological Economics, 150*: 264–272.

Kutty, N. (2022). Why Japan Sees Regeneration as Key to a Successful Circular Economy, *World Economic Forum*, 8 November. www.weforum.org/agenda/2022/11/is-regeneration-the-key-to-the-future-of-the-circular-economy/?DAG=3&gclid=CjwKCAiAh9qdBhAOEiwAvxIok7TvUW7

0M0L3j8RjuKRBqeFTic9nYJqlRFC8T9UZ6NX1F86IgyBr7hoCcEIQ AvD_BwE. Accessed 8 July 2024.

LaPensee, E.W., Tuttle, T.R., Fox, S.R., and Ben-Jonathan, N. (2009). Bisphenol A at Low Nanomolar Doses Confers Chemoresistance in Estrogen Receptor-Alpha-Positive and -Negative Breast Cancer Cells, *Environmental Health Perspectives*, *117 (2)*: 175–180.

Lepawsky, J. (2018). *Reassembling Rubbish: Worlding Electronic Waste*. MIT Press.

Levis, J.W., Barlaz, M.A., Themelis, N.J., and Ulloa, P. (2010) Assessment of the State of Food Waste Treatment in the United States and Canada, *Waste Management*, *30 (8–9)*: 1486–1494.

Lougheed, S. (2017) *Disposing of Risk: The Biopolitics of Recalled Food and the (Un) Making of Waste*. PhD thesis. https://qspace.library.queensu.ca/bitstream/ handle/1974/23777/Lougheed_Scott_C_201712_PhD.pdf?sequence=3. Accessed 19 October 2024.

Lougheed, S., Hird, M.J., and Rowe, R.K. (2016) Governing Household Waste Management: An Empirical Analysis and Critique, *Environmental Values*, *25 (3)*: 287–308.

MacBride, S. (2012). *Recycling Reconsidered*. MIT Press.

Magdoff, F. and Williams, C. (2017). *Creating an Ecological Society: Toward a Revolutionary Transformation*. Grassroots Books.

Mah, A. (2021). Future-Proofing Capitalism: The Paradox of the Circular Economy for Plastics, *Global Environmental Politics*, *21 (2)*: 121–142. Marscheider-Weidemann, F., Langkau, S., Baur, S.-J., Billaud, M., Deubzer, O., Eberling, E. et al. (2021). *Raw Materials for Emerging Technologies 2021 Report*. DERA Rohstoffinformationen 50. www.deutsche-rohstoffagentur.de/DE/Gemeinsames/Produkte/Downloads/DERA_Rohstoffinformationen/rohstoffinformationen-50-en.pdf?__blob=publicationFile&v=2. Accessed 8 July 2024.

Marten, B. and Hicks, A. (2018) Expanded Polystyrene Life Cycle Analysis Literature Review: An Analysis for Different Disposal Scenarios, *Sustainability*, *11 (1)*: 29–35.

Michaux, S.P. (2021). *Assessment of the Extra Capacity Required of Alternative Energy Electrical Power Systems to Completely Replace Fossil Fuels*. GTK Open Work File Internal Report, Serial number 42/2021.https://tupa.gtk.fi/ raportti/arkisto/42_2021.pdf. Accessed 8 July 2024.

Mothiba, M., Moja, S.J., and Loans, C. (2017). A Review of the Working and Health Conditions of Waste Pickers at Some Landfills in the City of Tshwane Metropolitan Municipality, South Africa. *Advances in Applied Science Research*, *8*: 90–97.

Murray, A. (2019) The Incinerator and the Ski Slope Tackling Waste, *BBC News*, *4 October*. www.bbc.com/news/business-49877318. Accessed 10 May 2021.

National Geographic. (2018). 7 Things You Didn't Know About Plastic and Recycling. https://news.nationalgeographic.org/7-things-you-didnt-know -about-plastic-and-recycling/. Accessed 27 October 2024.

Olmer, N., Comer, B., Roy, B., Mao, X., and Rutherford, D. (2017) Greenhouse Gas Emissions from Global Shipping, 2013–2015, *International Council on Clean Transportation*, 17 October. https://theicct.org/publications/GHG-emissions-global-shipping-2013–2015. Accessed 14 May 2021.

O'Neill, K. (2019). *Waste*. Polity Press.

Parizeau, K. (2006). A World of Trash: From Canada to Cambodia, Waste Is a Common Problem with Common Solutions. *Alternatives Journal, 32* (*1*): 16–19.

PlasticsEurope. (2018). *Plastics – The Facts 2018*. PlasticsEurope AISBL.

Preisner, M., Neverova-Dziopak, E., and Kowalewski, Z. (2020). 'An Analytical Review of Different Approaches to Wastewater Discharge Standards with Particular Emphasis on Nutrients', *Environmental Management, 66*: 694–708.

Raworth, K. (2017). *Doughnut Economics: Seven Ways to Think Like a 21st-Century Economist*. Chelsea Green Publishing.

Rowe, R.K. (2012). Third Indian Geotechnical Society: Ferroco Terzaghi Oration Design and Construction of Barrier Systems to Minimize Environmental Impacts Due to Municipal Solid Waste Leachate and Gas, *Indian Geotechnical Journal, 42* (*4*): 223–256.

Savini, F. (2019). The Economy That Runs on Waste: Accumulation in the Circular City, *Journal of Environmental Policy and Planning*. https://doi.org/10.1080/1523908X.2019.1670048.

Sayer, A. (2000). Moral Economy and Political Economy, *Studies in Political Economy, 61*: 79–103.

Skill, K. (2008) *(Re)Creating Ecological Action Space: Householders' Activities for Sustainable Development in Sweden*. PhD thesis, Linköping University.

Stål, H.I. and Corvellec, H. (2018). A Decoupling Perspective on Circular Business Model Implementation: Illustrations from Swedish Apparel, *Journal of Cleaner Production, 171*: 630–643.

Takai, Y., Tsutsumi, O., Ikezuki, Y., Hiroi, H., Osuga, Y., Momoeda, M., Yano, T., and Taketani, Y. (2000) Estrogen Recipient-Mediated Effects of a Xenoestrogen, Bisphenol A on Preimplantation Mouse Embryos, *Biochemistry and Biophysical Research Communications, 270* (*3*): 918–921.

Thompson, J. and Anthony, H. (2008) The Health Effects of Waste Incineration, in *Fourth Report of the British Society for Ecological Medicine* (2nd edn). www.ecomed.org.uk/content/IncineratorReportv3.pdf. Accessed 2 October 2012.

UNEA UNEP (United Nations Environment Assembly of the United Nations Environmental Programme). (2022). *End Plastic Pollution: Towards an International Legally Binding Instrument*. https://wedocs.unep.org/bitstream/handle/20.500.11822/38522/k2200647_-_unep-ea-5-l-23-rev-1_-_advance.pdf?sequence=1&isAllowed=y.4. Accessed 19 October 2024.

UNECE. (1999). *Protocol to the 1979 Convention on Long-range Transboundary Air Pollution to Abate Acidification, Eutrophication and Ground-level Ozone*. (Gothenburg Protocol). UNECE.

UNEP (United Nations Environmental Programme). (2024). *Bend the Trend: Pathways to a Liveable Planet as Resource Use Spikes*. www.resourcepanel.org/sites/default/files/documents/document/media/gro24_full_report_1mar_final_for_web.pdf. Accessed 8 July 2024.

United Nations. (1997). *Kyoto Protocol to the United Nations Framework Convention on Climate Change*. UN.

Vendries, J., Sauer, B., Hawkins, T.R., Allaway, D., Canepa, P., Rivin, J., and Mistry, M. (2020). 'The Significance of Environmental Attributes as Indicators of the Life Cycle Environmental Impacts of Packaging and Food Service Ware', *Environmental Science and Technology, 54*: 5356–5364.

Völker, T., Kovacic, Z., and Strand, R. (2020). Indicator Development as a Site of Collective Imagination? The Case of European Commission Policies on the Circular Economy, *Culture and Organization, 26* (*2*), 103–120. https://doi.org/10.1080/14759551.2019.1699092.

Vonk, L. (2018). Paying Attention to Waste: Apple's Circular Economy, *Continuum, 32* (*6*): 745–757.

Wheeler, K. and Glucksmann, M. (2015) 'It's Kind of Saving Them a Job Isn't It?' The Consumption Work of Household Recycling, *Sociological Review, 63* (*3*): 551–569.

Wilson, D.C. and Velis, C.A. (2015). Waste Management: Still a Global Challenge in the 21st Century; an Evidence-Based Call for Action, *Waste Management and Research, 33* (*12*): 1049–1051.

Yang, S. and Furedy, C. (1993). Recovery of Wastes for Recycling in Beijing, *Environmental Conservation, 20*: 79–82.

CONTEXTUALIZING WASTE

OVERVIEW

Chapter 5 focuses on waste as a social injustice issue. Indigenous and activist groups, non-governmental organizations such as Greenpeace and Human Rights Watch, and an increasing number of waste studies researchers are exposing the links between waste and economic, political, and social injustice. Indigenous, racialized, and income-insecure communities around the world are more likely to be on the front lines of waste issues because of their closer proximity to toxic waste sites. Waste exports from wealthy to poor regions of the world is a social injustice issue because wealthy regions externalize the human health and pollution problems to more vulnerable people. Waste picker jobs are precarious and expose children, women, and men to daily physical hazards for very little pay. The ongoing creation of waste landscapes through historical and contemporary colonialism is increasingly acknowledged. Illuminating waste as a social injustice issue highlights that dump, landfill, and incinerator siting, waste exporting and a host of other waste-related practices are differentially organized worldwide by powerful private corporate, industry, and government interests. This examination highlights the need to reject waste as a problem of consumer responsibility alone, and to engage with waste as a social injustice issue.

DOI: 10.4324/9781003398424-5

INTRODUCTION

This chapter extends Chapter 3's outline of waste as a social injustice issue. Private companies and governments largely present waste management as a combination of the need for more innovative technologies, and consumer responsibility. When discussions of waste are isolated from profound forms of inequality, then it is far easier to maintain the focus on individual responsibility and techno-management. Without a social injustice perspective, it is difficult to see why waste is a complex problem that is unresolvable through individual choices alone. In other words, solving our global waste crisis is impossible through individual initiatives such as more and better recycling, green consumerism, beach clean-ups, and the like. Local, national, and international non-governmental organizations such as Greenpeace and MiningWatch are critical in reporting on human rights and environmental abuses from extraction, manufacturing, and retail industries, and putting waste into global perspective. Chapter 5 focuses on communities facing such issues as the contaminating effects of leaking landfills, polluting energy-from-waste facilities and mining operations, and the impacts of waste exports to impoverished regions. It also examines the stigmatized and highly precarious occupation of waste picking as one of the most vulnerable occupations that regularly incurs negative health impacts.

ENVIRONMENTAL RACISM

The World Economic Forum describes environmental racism as a:

> form of systemic racism whereby communities of color are disproportionately burdened with health hazards through policies and practices that lead them to live in proximity to sources of toxic waste such as sewage works, mines, landfills, power stations, major roads and emitters of airborne particulate matter.
>
> (2020)

The children, women, and men in these communities face greater negative health impacts, and the surrounding environment is negatively affected by pollution. In the United States, as elsewhere, commercial hazardous waste facilities are far more likely to be sited near racialized and income-insecure communities (Mohai and Saha 2015). This said, poverty may not be the determining factor. Studies demonstrate that race is more important than poverty per se: racialized people who have similar income levels to non-racialized people are still more likely to be exposed to industrial pollution and to the negative impacts of climate change (Mothiba, Moja, and Loans 2017).

Environmental racism takes myriad forms. In many cases, environmental racism takes the form of negative human health and environmental impacts when open dumps, landfills, and waste-producing facilities are situated in or close to racialized communities. In other cases, waste pickers experience the toxic effects of waste exports from other regions and countries. And in still other cases, environmental racism is endemic to neo-colonial control of Indigenous land for the purposes of resource extraction, munitions testing, and other activities that produce contaminated waste sites.

WASTE'S NEGATIVE HUMAN HEALTH AND ENVIRONMENTAL IMPACTS

Benjamin Chavis's early identification of environmental racism in 1982 (see Chapter 3) in the case of the polychlorinated biphenyl waste that was disproportionately affecting a predominantly Black community situated near a landfill in Warren County, North Carolina, helped to launch similar investigations into the relationship between toxic waste contamination and vulnerable racialized communities (Chavis and Lee 1987). In 2014, Flint Michigan was finally investigated when residents' long-standing complaints about foul-smelling and clouded water, hair loss, and skin rashes resulted in between 6,000 and 12,000 children being exposed to unacceptable levels of lead (a neurotoxin) and twelve people dying from Legionnaires' disease (Denchak 2024). The problem turned out to be the City's shift to sourcing the municipality's drinking supply from the polluted Flint River and failing to adequately treat the water supply before human consumption. When the case finally made its way to court, the Michigan Civil Rights Commission found that the long delay in official reaction and remediation was the 'result of systemic racism' (2017: iii). People living in Louisiana's 'Cancer Alley', that stretches along the Mississippi River from New Orleans to Baton Rouge are 50 times more likely to suffer from cancer and other health effects produced by the oil refin-eries and petrochemical plants clustered in this area (Pasley 2020). Flint, New Orleans, and Baton Rouge join Detroit, Michigan, War-ren County, North Carolina, Houston, Texas, Los Angeles, California, The Bronx, New York City, Cheraw, South Carolina, Uniontown, Alabama, and Pahokee, Florida in disproportionately exposing racial-ized communities to toxic waste (Colarossi 2020).

We find numerous examples of the disproportionate effects of waste toxicity and contamination on racialized and income-insecure communities throughout the world, and also within afflu-ent countries (see Hird 2021 and Human Rights Watch 2020 for other examples). Often, the negative human health and environ-mental impacts of waste are not immediately known and take years to come to public attention. And even when cases do come into public view, they are not necessarily resolved. As such, there are many more cases of environmental racism than those that have been brought to the public's attention.

EXPORTING WASTE

On 31 August 1986, the Liberian cargo ship Khian Sea was loaded with over 14,000 tons of incineration ash waste from Philadelphia, Pennsylvania. Exports, including waste, often change company hands several times between point of origin and final destination, and Joseph Paolino and Sons, the company contracted by Philadelphia, subcontracted the shipment to Amalgamated Shipping Corp and Coastal Carrier Inc, who intended to offload the waste shipment in the Bahamas. When the Bahamian government refused to allow the cargo ship to dock, it triggered a more than two-year saga. The subcontractors and the crew of the Khian Sea attempted to offload the waste in several countries, including Honduras, Panama, Bermuda, Guinea Bissau, the Dutch Antilles, and the Dominican Republic (Leonard 2011).

Everyone refused to accept the waste. The subcontractors even tried to return the waste to its origin, Philadelphia, where officials also refused to repatriate its incineration garbage. Then, in January 1988, the crew dumped some 4,000 tons of the toxic waste in Haiti, calling it 'topsoil fertilizer' (Reeves 2001). Greenpeace alerted the Haitian government to the illegal dumping, but the ship left before government officials could compel the crew to reload the waste. While some of the waste was sequestered in a bunker, most of it remained on the beach, open to the environmental elements. The crew attempted to offload the remainder of its load in Sri Lanka, Singapore, Morocco, Yugoslavia, and Senegal, without success. The ship even changed its name from the Khian Sea to Felicia and then to Pelicano in an attempt to disguise its waste load. Finally, in 1988, over two years after the cargo ship left the United States, the ship's captain and crew dumped the ash waste in the Atlantic and Indian Oceans. In 1997, the New York City Trade Waste Commission agreed to give Eastern Environmental Services a New York operating license on condition that it contribute to the cleanup in Haiti. The city of Philadelphia contributed $50,000 to the cleanup operation. In 2000, Waste Management Inc., one of the world's biggest multinational waste management corporations, shipped some 2,500 tons of the toxic ash and contaminated soil to Florida, where it remained on the ship for two years before being finally offloaded to a landfill in Pennsylvania. The long and very public saga of this orphaned waste shipment contributed to the initiation of the Basel Convention's transboundary waste export regulations (Leonard 2011; see Chapter 4).

In 2020, Imani Williams remarked, 'One man's trash is another man's burden'. Exporting waste to other regions and countries is a widespread practice. A growing exposé of waste exports from wealthy countries to impoverished countries illuminates the profound negative health impacts on the communities that are on the receiving end of waste. The Khian Sea fiasco is one example of a disturbingly common phenomenon of wealthy countries offloading their waste (and its problems) onto poorer countries. The European Union is the largest exporter of waste, followed by the United States as the largest single country exporter (BBC News 2019). Greenpeace UK (2021) recently exposed the United Kingdom (ranked with the US as the world's leading plastics waste producer) for exporting plastics waste to poor regions of the world, despite government claims to the contrary. In 2020 alone, the UK exported 688,000 tons of plastics packaging to Turkey, Malaysia, Poland, and other countries, while recycling less than half of that within its own borders. Greenpeace Aotearoa revealed that New Zealand offloaded 98,000 tons of plastics waste to Malaysia, Thailand, and other poorer countries between 2018 and 2021 (Morunga 2021). These countries are ill-equipped to deal with the increasing mountains of plastics and other waste, which is often burned in open dumps in the vicinity of neighborhoods, schools, and other residential spaces.

Before 2018, China imported the greatest percentage of plastics and other recycling (such as copper, aluminum, and paper). In 2016 alone, the United States exported 1,500 shipping containers of plastics and other scrap (such as paper and metals) *per day* to China (Flower 2016). While China had already attempted to limit scrap imports before, most notably through its Operation Green Fence in 2013, in 2017 China very publicly declared an import cessation – calling it the National Sword Policy – sending many countries into a panicked tailspin. Malaysia largely picked up the waste exports that would have gone to China, and like other countries that import waste such as Vietnam, South Korea, India, Taiwan, Indonesia, and Thailand, it is now experiencing the same kinds of issues that China faced.

Much of this ever-increasing recycling is actually contaminated and cannot be recycled: that is, much of the plastics labeled as recycling at the point of origin is actually dirty plastics that have been mixed with regular non-plastics garbage (see Chapter 4). And this means that the receiving countries cannot glean any profit from recycling and are burdened with increasing waste disposal problems. Many countries are scrambling to find ways of continuing this linear waste export chain as well as deal with an increasingly informed and discontented public demanding change.

WASTE PICKERS

As Chapter 1 describes, an increasing number of the world's population live either on or in close proximity to open dumps, and mega slums are increasingly common. Globally, some 20 million people sustain themselves and their families as waste pickers (Hillsdon 2023). According to the organization Women in Informal Employment: Globalizing and Organizing (WIEGO), this is likely a significant underestimation, as many waste pickers either live on the dumps at which they work or on the street, and are thus not included in household surveys. Moreover, because waste picking tends to be highly stigmatized (and sometimes criminalized), many waste pickers do not want to be included in population surveys. Waste pickers are almost always part of the informal economies of low-income and middle-income countries, supporting either no or inadequate waste disposal and recycling infrastructure and systems. Waste pickers are essentially the world's lowest paid sanitation workers and recyclers.

Waste pickers gather together and sort through colossal piles of discarded materials on open dumps or landfills. They collect waste from households and public spaces, and demolish buildings with often little equipment (sledgehammers and bare hands) and no protective gear. Waste pickers include people who are sorting through waste sites or moving material to waste sites, rivers, or coastal waters. In some regions, informal waste pickers provide the only form of solid waste collection. The Global Review on Safer End of Engineered Life notes that waste pickers remove over 88 million tons of materials such as plastics, glass, metal, paper, and cardboard for recycling each year (Cook and Velis 2021).

Waste picking is one of the most hazardous, low-paying, and precarious jobs in the world. There are numerous safety concerns when people are picking through open dumps or manually demolishing buildings, from the spontaneous combustion of waste, to falling and/or toxic debris, to open dump landslides. Many materials are disposed of in open dumps, such as medical waste, which may contain blood, urine, feces, body parts, and other biohazardous materials, as well as various pathogens. There are also sharp objects (which may or may not be contaminated) that can lead to infections, especially when many waste pickers do not have access to protective equipment such as gloves (Cook et al. 2020a). Medical waste is recovered for sale to mainstream health care providers in low- and middle-income countries as well as for use by substance users. Electronic waste recovery involves separating metals such as copper and components from laptops, smart phones, or other devices through heating, combustion, and/or the use of acids and alkali, all of which expose waste pickers to inhaling and touching hazardous substances such as lead, chromium and arsenic, brominated flame retardants, and other potentially toxic elements (Cook et al. 2020b). Several studies have found elevated levels of lead in pre-school child waste pickers in China (Lin et al. 2017; Lin et al. 2016). Plastics waste burning further exposes waste pickers to over 16,000 different chemicals, some 4,000 or more of which are hazardous (Jones 2024). Construction waste includes toxic materials such as asbestos and gypsum. Asbestos alone is reported to have caused the deaths of 233,000 workers worldwide (Furuya et al. 2018). And whereas some wealthy countries such as Canada have banned the production and use of asbestos, some thirty-nine countries still produce 1.1 million tons per year (National Minerals Information Center 2018; Cook, Velis, and Black 2022), and some 1.25 million people are expected to develop cancer from asbestos inhalation (Jadhav and Gawde 2019).

COLONIALISM AND WASTE

The global neoliberal capitalist system structures waste's governance and management such that private companies benefit from favorable, lax, or non-existent regulations that would prevent or effectively limit resource extraction, manufacturing, and distribution waste, and such that individual responsibility for post-consumption waste is disproportionately amplified. Lorenzo Veracini distinguishes between colonialism and settler colonialism: 'whereas colonialism reinforces the distinction between colony and metropole, settler colonialism erases it' (2011: 3). The history of colonialism and settler colonialism is beyond the scope of this book (see Veracini 2010, 2011; Yusoff 2018). The nature of settler/colonialism (to include both) is to make territory available for the colonizing government's profit through resource extraction and manufacturing (using cheap and slave labor).

Capitalism would not have been possible in either its global reach or effect without historical and ongoing (settler) colonialism. As Kathryn Yusoff observes, affluent nations' recent concern with exposure to environmental harms comes 'in the wake of histories in which these harms have been knowingly exported to black and brown communities under the rubric of civilization, progress, modernization, and capitalism' (2018: 11). The intricate undergirding of capitalism by colonialism is beyond the scope of this book (see Wynter n.d.; James 1938/1989; Malm 2015; Malm and the Zetkin Collective 2021). As it relates specifically to waste, one of the major aims of colonialism is to secure ongoing access to a colonized territory's natural resources. Countries in Africa – Ghana, Angola, Zimbabwe, South Africa, Mali, Botswana, Democratic Republic of the

Congo, Namibia, Burkina Faso, and others – are all surviving histories of colonialism that violently subjugated peoples for either slave or cheap labor, and continue to use this labor for minerals extraction. François Jarrige and Thomas Le Roux refer to the Democratic Republic of the Congo as the 'archetype of a colonized country pillaged for its mineral wealth and the continuity of that exploitation after decolonization' (2020: 282–283). The researchers note that Belgian, British, and American companies have extracted gold, diamonds, manganese, uranium, tungsten, tin, tantalum, copper, and cobalt from the Katanga region, using compromised local labor under extreme conditions of hardship, and devastating the environment.

Canada/Turtle Island, the United States, Australia, New Zealand/Aotearoa, New Caledonia, and other oil- and mineral-rich territories have violent histories of settler colonialism that extracted (and continue to extract) precious metals, oil, gas, and other valuable resources from unceded Indigenous territories (Hird and Predko 2024). In Canada, for instance, there are approximately (the exact number is unknown) 27,000 abandoned or 'orphaned' mines, most of which are in Canada's northern Indigenous regions. The Giant Mine, located on the Ingraham Trail close to Yellowknife, was abandoned in 2005, leaving some 100 on-site buildings, eight open pits, and contaminated soils and waste rock around the mine, including some 237,000 tons of arsenic trioxide dust (Sandlos and Keeling 2012). The Giant Mine is situated on Wıìlıìdeh Yellowknives Dene First Nation, and this Nation is now burdened with a perpetual toxic contamination that they did not create. There are over 10,000 orphaned oil and gas wells in Alberta alone (Orphan Well Association 2023). In New Caledonia, where the country contains some 7.1 million tons of nickel (about ten percent of the world's total nickel reserves), strip mining has produced significant environmental and health impacts. While most of the front-line miners are local Indigenous Kanak people, the French company Eramet owns over fifty percent of the nickel industry. Sandwiched uncomfortably between an independent country and a 'collectivité', France maintains a strong grip on New Caledonia in order to continue to secure its resource extraction dividends. The Kanak people's increasing efforts to decolonize their land has recently led to increasingly violent protests, with French President Emmanuel

Macron visiting the island in May 2024 in an attempt to quell these independence efforts (Derel and Chrisafis 2024). Canada and New Caledonia's conflicts with Indigenous peoples over land and water rights are mirrored throughout the world where colonizing countries continue to occupy or otherwise control Indigenous land (Scheidel et al. 2020).

There are no signs that resource extraction is abating: in fact, the opposite is true. The Arctic States – Canada, Greenland, Iceland, Norway, Sweden, Russian, Finland, and the United States – are in an increasingly contentious negotiation to secure Arctic land and waters for the purpose of resource extraction. Each nation knows that the potential economic dividends are great: whichever country secures a claim stands to vastly increase its country's access to diminishing and increasingly valuable natural resources, such as oil and gas. The Canadian federal government, for instance, claims that about twenty-five percent of Canada's remaining crude oil and natural gas and about forty percent of its projected future discoveries are to be found in the Arctic (Government of Canada 2010). As the next chapter details, the increasingly urgent calls to decarbonize the transportation sectors through electric battery vehicles is likely to lead to further (and potentially more intense) human rights violations, environmental harm, and social injustice unless governments rein in mining corporations and companies through strict regulations in tandem with significant changes to current transportation infrastructures (Riofrancos et al. 2023).

Colonizing nations also use colonized territories in ways that produce extreme contamination – contamination that would not be tolerated within the colonizing nation's home territory. Nuclear bomb testing is a primary example of the impact of colonialism on toxic contamination. The Republic of the Marshall Islands is a United States associated state and comprises some 1,156 islands in the Pacific Ocean, north of New Zealand. With a total population of just over 58,000, most of its territory (over ninety-seven per cent) is water. Beginning around the tenth century, successive waves of colonizers and settler colonizers claimed the islands, from Micronesians, to Spanish, to Germans, to Japanese, and finally to Americans during World War II. The US began nuclear bomb testing on the Marshall Islands' Bikini Atoll in 1946 and continued detonating nuclear arms for over a decade. During this period, the US exploded twenty-three nuclear weapons, first above-ground and then underground. The second – Baker test – detonation contaminated all of the surrounding ships, leading Glenn T. Seaborg, chair of the Atomic Energy Commission, to call it 'the world's first nuclear disaster' (in Weisgall 1994: ix). As well as widespread lasting nuclear contamination affecting generations of islanders, flora, and fauna, this military operation also left on-site ninety-five navy ships, including cruisers, destroyers, submarines, attack transports, landing ships, carriers, and battleships, all carrying fuel and some with live ammunition. And it also left the Runit Dome, known locally as 'The Tomb', which is a bomb crater made from the eighteen kt Cactus detonation in 1958. As Peter van Wyck details:

> Between 1977 and 1980, contaminated topsoil and debris from the atoll (including 16,000 items of WWII ordinance, such as unexploded artillery projectiles, mortar shells, hand grenades, and small arms ammunition) were bulldozed into the crater, and a 45 cm concrete cap, or dome, was constructed on the surface. The Dome is now at risk of failure from deterioration, saltwater incursion, vulnerability to typhoons and sea level rise.
>
> (in Kavanagh 2020: 49)

Since the first nuclear detonation on July 15th, 1945, there have been some 2,056 nuclear detonations across the globe, most of them exploded on colonized Indigenous lands far away from the capitals of colonizing forces (Arms Control Association 2020). France, for instance, tested its first nuclear weapon, code-named 'Gerboise Bleue' (Blue Desert Rat), in Algeria in February 1960, which France had violently invaded and then colonized. This first French nuclear test was recorded at 70 kt, or as powerful as the US bomb dropped on Hiroshima, Japan, during World War II. France also made use of some of its other colonies in the French Polynesian atolls in the South Pacific as well as controversially conducting the last of its 210 nuclear tests there in 1996, during the Comprehensive Nuclear Test-Ban Treaty (CTBT) negotiations in Geneva. France signed on to the treaty only after international protests included French export boycotts. Only in 2009 did France's Senate acknowledge the impacts of its testing program and provide some compensation to civilian and military veterans.

France's nuclear testing had carcinogenic and other negative health effects on local residents: atmospheric plutonium-239 concentrations were found to be four times greater in these French colonies than in continental France, leading in some cases to the evacuation of whole islands. It also resulted in considerable damage to the environment. Radiation from the blasts led to declines in livestock and biodiversity. And, like the Runit Dome, France stored radioactive waste (including plastic bags, clothing, metal scrap, and wood) on the north coast of the colonized Mururoa atoll, in an area covering about 30,000 square meters. When cyclones hit Mururoa in 1981, radioactive waste was washed into the lagoon, including the plutonium-239-laden bitumen that had been used to contain the plutonium.

COMMUNITY ACTIVISM FOR WASTE JUSTICE

Indigenous peoples around the globe are often on the front lines of efforts to draw attention to resource extraction and manufacturing's negative human health and environmental impacts. They are on these front lines because a lot of industrial activity takes place where Indigenous peoples live, work, and raise their families. Anishinaabekwe leader and Harvard graduate Winona LaDuke (2016), for instance, chronicles the efforts of the White Earth Land Recovery Project, the Indigenous Women's Network, and thousands of Indigenous people and allies in protesting the Northern Gateway, Keystone, and Enbridge Great Lakes Pipelines that would move oil from Alberta, Canada to the United States, traversing wide swaths of traditional Sioux, Ojibwe, and other First Nations territory. Indigenous peoples, as well as income-insecure remote communities, are far more likely to be in close proximity to oil, gas, and mining operations and other industrial zones, and thus directly experience the negative effects of waste.

For example, the Kitchenuhmaykoosib Inninuwug (KI) of North Western Ontario, Canada challenged the mining company Platinex's plans for mineral exploration on traditional KI land and water, doing so on the argument that their culture is based on an intimate relationship (involving history and identity) with land (Ariss 2017; Ariss and Cutfeet 2011). In 2008, six leaders of the KI community were imprisoned for peacefully protesting resource development on their land. The KI-6, as they became known, were released two months into their sentence. The KI community challenged the Ontario courts, successfully arguing that Platinex's plans violated the community's rights under Treaty 9. As the KI-6 repeatedly reminded government and industry officials, their obligation to protect their environment extended to all things (for instance, the fish in Big Trout Lake) and people, not just the KI community itself. Similarly, the Qamani'tuaq (Baker Lake) community in the Kivalliq Region of Nunavut challenged French multinational Areva's proposal to construct a uranium mine and store radioactive waste 80 kilometers west of their community (Metuzals and Hird 2018; Kuyek 2019). And in the wake of Brazil's Córrego de Feijão iron mine waste tailings pond rupture in 2019, which led to an environmental devastation some several hundred kilometers long through five Brazilian states and the death of at least 209 local inhabitants, local residents and activist groups as well as international organizations such as Greenpeace are – despite threats and harassment – attempting to hold Vale Canada Limited accountable (Wilkes and Hird 2019). Martín Arboleda reminds us that 'the immanent dynamics that underpin the spaces of extraction of late capitalism are global in content and national only in form' (2020: 26). Often those communities that are most devastated by what we might call the 'fast violence' of environmental disaster, and the 'slow violence' (Nixon 2011; see also Hume and Tawfeeq 2016) of environmental degradation, are also on the front lines of challenging highly enfranchised, and government protected, companies and corporations, with often the least financial and other resources.

When we understand the links between social injustice and waste, we see that waste is political. While North Korea's 'trash balloons' sent to South Korea may be an obvious political stunt, many products today 'hid[e] subtle pollutions incorporated into the heart of the production process, from the mine to final consumption' (Jarrige and Le Roux 2020: 231; Yeung and Seo 2024). Numerous companies work hard to obscure the environmental and human health costs of the products they manufacture and sell, as they also work hard to disguise the labor conditions of their workers. Thus, it is often the case that only local residents, whose communities are close to resource extraction and manufacturing sites, and/or whose labor makes the products, are aware of the negative impacts. Cancer alleys, mega dumps, superfund sites, and sacrifice zones are intimately known to the people who live close by, and remain literally and figuratively remote to most other people. The environmental costs of affluent countries' overconsumption have been, up until relatively recently, kept mostly beyond the view of consumers, as are the sites of extraction and production. This is changing. People in affluent regions are increasingly aware that waste is a global crisis of social injustice.

NON-GOVERNMENTAL ORGANIZATIONS FOR WASTE JUSTICE

As well as grass-roots community organizations protesting waste issues throughout the world, there are a wide range of non-governmental organizations that support communities in advocating for waste reduction and what we might call 'waste justice'. These non-governmental organizations issue reports about particular waste issues, lobby governments to enact, regulate, and enforce stricter legislation to control and reduce waste and pollution, and raise awareness through publicly accessible publications and advertising (via websites and social media) about both local and global human health and environmental concerns caused by waste.

One critical feature of non-governmental organizations that requires attention is their funding sources. Some organizations, such as Greenpeace, rely solely on individual donations, and can therefore act independently of interest and lobby groups that have a stake in presenting their industry in a favorable way. They are also transparent in reporting who donates to their organization. Other organizations, such as PlasticsEurope, are trade organizations for industry. This means that their publications and advertising favor industry-friendly messaging, including how waste problems are understood, and what solutions to these problems are highlighted. On its website, PlasticsEurope states 'We are increasing our efforts to reduce plastics waste, promote the reuse and collection and recycling of plastics waste, and have accelerated the transition to a circular economy in response. We are striving to achieve 'zero plastics to landfill' and 100% recovery of plastics waste' (PlasticsEurope.org). This plastics-industry-supported organization repeats the fossil fuel and plastics industry claims that recycling effectively deals with plastics waste, and that a Circular Economy is possible (for a critique of both of these claims, see Chapter 4).

Another issue facing non–governmental organizations that accept monetary and in-kind support from companies is that they may inadvertently provide positive advertising for these companies, and thus undermine their own goals. For example, the Human Nature Projects (HNP) Ontario advertises itself as a 'federally registered, youth-led, non-profit organization seeking to raise awareness for current environmental issues' (hnpontario.org). Its goal 'is to drive environmental change in Ontario and help educate Ontarians about pressing environmental concerns'. Their events include community clean-ups at various parks, invasive species removal, and tree planting. Their website states that in 2022 over 170 community volunteers helped collect eighty bags of garbage from parks and over 250 volunteers collected over thirty bags of garbage from ponds. The organization even explored the theme of greenwashing through a liberal arts competition.

However, HNP is sponsored by a number of companies, including EcoSpark. While EcoSpark is a registered charity, it is sponsored by the Royal Bank of Canada (RBC), which is the leading Canadian banking company investing in the fossil fuel industry (Meyer 2024). The Wet'suwet'en land defenders and other Indigenous First Nations have been protesting the construction of fossil fuel pipelines – supported by the RBC – on their unceded territory without their consent. EcoSpark is also sponsored by Enbridge, a multinational fossil fuel pipeline and energy company. Founded by Imperial Oil, Enbridge is responsible for over 27,564 kilometers of pipelines that cross Indigenous lands throughout Canada and the United States, transporting more than three billion barrels of oil and liquids *per day*, making it the world's longest crude oil pipeline (CBC News 2020; see also www.enbridge.com). One of HNP's major activities is community clean-up at which volunteers collect waste in public spaces. As Chapter 3 details, beach and other community cleanups serve companies well. Companies and corporations such as Enbridge and the RBC are able to include their sponsoring of EcoSpark and HNP in their financial reporting. And more importantly, it allows these companies and corporations to publicize that they are helping the environment, helping young people to become 'environmental citizens', and so on as part of their commitment to 'corporate responsibility'. These cleanups do nothing to stop the production of waste, and effectively divert attention away from waste production. They, as Chapter 3 argues, do nothing to 'turn off the tap' of industry waste production. Thus, inadvertently, some non-governmental organizations enable the very kinds of waste production and greenwashing that they are attempting to challenge. This is not to condemn the efforts of the volunteers themselves, but to ask critical questions about the consequences of greenwashing and individual responsibility 'solutions' to waste that serve the interests of polluting industries, create 'busy work' that deflects attention from industry responsibility for waste, and does nothing to reduce the sources of waste (see Chapters 3 and 4).

Numerous non-governmental organizations avoid compromising their goals by refusing financial support from industry. Chief among independent non-governmental organizations is Greenpeace. Greenpeace International was founded in 1971 in Canada by a group of environmental activists who began to campaign against the nuclear detonation tests carried out by the United States. Growing considerably since its modest beginnings, Greenpeace is now comprised of a network of twenty-six independent regional and national organizations in over fifty-five countries across the globe (Greenpeace.org). Several key Greenpeace publications directly concern waste, including the series on plastics pollution and the fossil fuel industry called 'Throwing Away the Future: How Companies Still Have it Wrong on Plastic Pollution "Solutions"' (2019). Greenpeace provides research information to supra-governmental organizations such as the United Nations. MiningWatch Canada is another independent organization that focuses specifically on mining waste and pollution issues concerning Indigenous peoples. MiningWatch lobbies governments and issues numerous reports detailing mining waste and other issues affecting Indigenous communities and land in Canada and around the world. Friends of the Earth focuses on issues related to climate change, including waste, such as reducing carbon emissions. Their transparent accountability and corporate donation policy helps to ensure that industry does not influence their platform. These and myriad other non-governmental organizations across the world are highlighting the unequal distribution of negative human health and environmental costs of waste and pollution. As such, these non-governmental organizations forefront waste as a social injustice problem much more than a problem of people not taking sufficient individual responsibility for their waste disposal and recycling. Understanding waste as social injustice is key to contextualizing our global waste crisis, getting to its root causes, and therefore devising solutions that will effectively resolve (rather than exacerbate) the problem.

REVIEW

Understanding the associations between waste and social injustice is critical to resolving waste problems at local and global levels. Within even affluent countries with modern waste management infrastructure, legislation, policies, and practices, people in low-income, racialized, and Indigenous communities are more likely to be exposed to pollution and other negative environmental waste (for example, Cancer Alley in Louisiana, United States). These communities may even have less access to otherwise standard municipal solid waste management (for example, Nunavut, Canada). Waste pickers are on the front lines of the effects of waste contamination, sorting through the waste of their own region as well as waste imported from wealthy countries. While international legislation such as the Basel Convention on the Transboundary Movement of Waste and China's National Sword Policy, as well as national rules and regulations, seek to end the illegal transportation of waste (most often in cases where waste is mislabeled as recyclable material), exporting waste remains a serious and difficult-to-resolve problem due to stark affluency differences between wealthy and impoverished nations, the latter of which often lack the power to refuse waste imports. Waste will remain a global crisis as long as producers as well as wealthy regions and nations are able to externalize the costs of waste.

SUGGESTED READING

Deniz, G.C. (2012). *Waste Pickers Who Carry the Weight of the World: A Case Study in Ankara on Waste Pickers and the Informal Waste Collection Sector*. LAP Lambert Academic Publishing.

LaDuke, W. (2017). *The Winona LaDuke Chronicles: Stories from the Front Lines in the Battle for Environmental Justice*. Fernwood Publishers.

O'Hare, P. (2022). *Rubbish Belongs to the Poor: Hygienic Enclosure and the Waste Commons*. Pluto Press.

Pellow, D.N. (2004). *Garbage Wars: The Struggle for Environmental Justice in Chicago*. The MIT Press.

REFERENCES

Arboleda, M. (2020). *Planetary Mine: Territories of Extraction Under Late Capitalism.* Verso Press.

Ariss, R. (2017). Platinex V Kitchenuhmaykoosib Inninuwug: Extraction and the Role of Law in KI's Struggle for Self-Determination, *Contours*, 7. https://jps.library.utoronto.ca/index.php/ilj/article/view/27637/20368. Accessed 27 October 2024.

Ariss, R. and Cutfeet, J. (2011). Kitchenuhmaykoosib Inninuwug First Nation: Mining, Consultation, Reconciliation and Law, *Indigenous Law Journal*, *10* (1): 1–37.

Arms Control Association. (2020). *The Nuclear Testing Tally.* www.armscontrol.org/factsheets/nucleartesttally. Accessed 31 December 2020.

BBC News. (2019) Why Some Countries Are Shipping Back Plastic Waste, *BBC*, 2 June. www.bbc.com/news/world-874. Accessed 20 May 2021.

CBC News. (2020). Enbridge Makes Deal to Sore Oil in Mainline Pipeline as Oil Glut Grows. *CBS*, 4 May. www.cbc.ca/news/canada/calgary/enbridge-mainline-1.5555509. Accessed 15 August 2024.

Chavis, B. and Lee, C. (1987) *Toxic Waste and Race in the United States: A National Report on the Racial and Socio-Economic Characteristics of Communities with Hazardous Waste.* United Church of Christ's Commission on Racial Justice.

Colarossi, N. (2020). 10 Egregious Examples of Environmental Racism in the US, *Business Insider*, 11 August. www.businessinsider.com/environmental-racism-examples-united-states-2020-8. Accessed 6 June 2024.

Cook, E. and Velis, C.A. (2021). *Global Review on Safer End of Engineered Life.* Report. Royal Academy of Engineering.

Cook, E., Velis, C.A., and Black, L. (2022). Construction and Demolition Waste Management: A Systematic Scoping Review of Risks to Occupational and Public Health, *Frontiers in Sustainability*, *3*. www.frontiersin.org/journals/sustainability/articles/10.3389/frsus.2022.924926/full. Accessed 27 October 2024.

Cook, E., Velis, C.A., Woolridge, A., Stapp, P., and Edmonson, S. (2020a). Medical and Healthcare Waste Generation, Storage, Treatment and Disposal: A Systematic Review of Risks to Occupational and Public Health, *Critical Reviews in Environmental Science and Technology*, *53*: 1452–1477.

Cook, E., Velis, C.A., Gerassimidous, S., Ramola, A., and Ragossing, A. (2020b) Physical Processing, Dismantling and Hydrometallurgical Treatment of E-waste: A Systematic Review of Risks to Occupational and Public Health, *Health and Medicine Week*, 25 December, 3890.

Denchak, M. (2024). *Flint Water Crisis: Everything You Need to Know.* Natural Resources Defense Council. www.nrdc.org/stories/flint-water-crisis-everything-you-need-know#sec-timeline.

Derel, M. and Chrisafis, A. (2024). Macron Vows Not to Rush Through Voting Change After New Caledonia Visit, *The Guardian*, 23 May. https://amp.the-guardian.com/world/article/2024/may/22/macron-arrives-in-new-caledo-nia-amid-calls-for-france-to-withdraw-voting-changes. Accessed 13 June 2024.

Flower, W. (2016). What Operation Green Fence Has Meant for Recycling, *Waste 360*, 11 February. www.waste360.com/business/what-operation-green-fence-has-meant-recycling. Accessed 10 February 2021.

Furuya, S., Chimed-Ochir, O., Takahashi, K., David, A., and Takala, J. (2018). Global Asbestos Disaster, *International Journal of Environmental Research and Public Health*, *15*: 1000.

Government of Canada. (2010). High Investment Potential in Canadian Northern Oil and Gas. *Indigenous and Northern Affairs Canada*. www.aadncaandc.gc.ca/eng/1100100037174/1100100037175.

Greenpeace. (2019). *Throwing Away the Future: How Companies Still Have it Wrong on Plastic Pollution 'Solutions'*. www.greenpeace.org/usa/research/how-companies-still-have-it-wrong-on-plastic-pollution-solutions/. Accessed 14 August 2024.

Greenpeace UK. (2021). *Trashed: How the UK Is Still Dumping Plastic Waste on the Rest of the World*. www.greenpeace.org.uk/wp-content/uploads/2021/05/Trashed-Greenpeace-plastics-report-final.pdf?_gl=1*1nu493u*_up*MQ.*_ga*MTY2MTg3MDc5NC4xNzE4MDE5MTQ2*_ga_94MR TN8HG4*MTcxODAxOTE0NS4xLjAuMTcxODAxOTE0NS4wLjAuM TA1ODgyNjkyMQ. Accessed 10 June 2024.

Hillsdon, M. (2023). How Waste Pickers Are Helping to Win the War on Plastics Pollution, *Reuters*, 17 July. www.reuters.com/sustainability/society-equity/how-waste-pickers-are-helping-win-war-plastic-pollution-2023-07-17/#:~:text=July%2010%20%2D%20There%20are%20an,whose%20work%20goes%20largely%20unrecognised. Accessed 6 June 2024.

Hird, M.J. (2021). *Canada's Waste Flows*. McGill-Queen's University Press.

Hird, M.J. and Predko, H. (2024). *Extracting Reconciliation: Inhuman Wastes, Indigenous Lands, and Colonial Reckoning*. Routledge.

Human Rights Watch. (2020) Lebanon: Huge Cost of Inaction in Trash Crisis, *HRW*, 9 June. www.hrw.org/news/2020/06/09/lebanon-huge-cost-inaction-trash-crisis. Accessed 1 January 2021.

Hume, T. and Tawfeeq, M. (2016) Lebanon: "River of Trash" Chokes Beirut Suburb as City's Garbage Crisis Continues, *CNN*, 25 February. www.cnn.com/2016/02/24/middleeast/lebanongarbage-crisis-river. Accessed 14 May 2021.

Jadhav, A.V. and Gawde, N.C. (2019). Current Asbestos Exposure and Future Need for Palliative Care in India, *Indian Journal of Palliative Care*, *25*: 587–591.

James, C.L.R. (1938/1989). *Black Jacobins: Toussaint L'Ouverture and the San Domingo Revolution*. Random House.

Jarrige, F. and Le Roux, T. (2020). *The Contamination of the Earth: A History of Pollutions in the Industrial Age.* The MIT Press.

Jones, N. (2024) More Than 4,000 Plastic Chemicals Are Hazardous, *Nature,* March 14.

Kavanagh, M. (2020). *Daughters of Uranium.* Southern Alberta Art Gallery; Founders' Gallery at The Military Museums; University of Calgary; and Kitchener-Waterloo Art Gallery.

Kuyek, J. (2019). *Unearthing Justice: How to Protect Your Community from the Mining Industry.* Between the Lines.

LaDuke, W. (2016). *The Winona LaDuke Chronicles: Stories from the Front Lines in the Battle for Environmental Justice.* Fernwood Publishing.

Leonard, A. (2011). *The Story of Stuff.* Simon and Schuster.

Lin, X., Xu, X., Zeng, X., Xu, L., Zeng, Z., and Huo, X. (2017). Decreased Vaccine Antibody Titers Following Exposure to Multiple Metals and Metalloids in E-Waste-Exposed Preschool Children, *Environmental Pollution, 220*: 354–363.

Lin, Y., Xu, X., Dai, Y., Zhang, Y., Li, W., and Huo, X. (2016). Considerable Decrease of Antibody Titers Against Measles, Mumps, and Rubella in Preschool Children from an E-Waste Recycling Area, *Science of The Total Environment, 573*: 760–766.

Malm, A. (2015). *Fossil Capitalism.* Verso.

Malm, A. and the Zetkin Collective. (2021). *White Skin, Black Fuel: On the Danger of Fossil Fascism.* Verso.

Metuzals, J. and Hird, M.J. (2018). 'The Disease that Knowledge Must Cure': Sites of Uncertainty in Arctic Development, *Arctic Yearbook.* https://arcticyearbook.com/arctic-yearbook/2018/2018-scholarly-papers/269-the-diseasethat-knowledge-must-cure-sites-of-uncertainty-in-arctic-development.

Meyer, C. (2024). Royal Bank of Canada on the Defensive Over Criticism of Fossil Fuel Financing, *The Narwhal,* 19 June. https://thenarwhal.ca/royal-bank-fossil-fuels-reputation/#:~:text=This%20year%2C%20Royal%20Bank%20was,International%20and%20the%20Sierra%20Club. Accessed 15 August 2024.

Michigan Civil Rights Commission. (2017). *The Flint Water Crisis: Systemic Racism Through the Lens of Flint.* 17 February. www.michigan.gov/-/media/Project/Websites/mdcr/mcrc/reports/2017/flint-crisis-report-edited.pdf?rev=4601519b3af345cfb9d468ae6ece9141. Accessed 27 October 2024.

Mohai, P. and Saha, R. (2015). Which Came First, People or Pollution? Assessing the Disparate Siting and Post-Siting Demographic Change Hypotheses of Environmental Injustice, *Environmental Research Letters, 10 (11)*: 115008.

Morunga, A.M. (2021). *The Devastation of New Zealand's Plastic Waste Exports on Developing Countries.* Greenpeace Aotearoa. www.greenpeace.org/aotearoa/

story/the-devastation-of-new-zealands-plastic-waste-exports-on-developing-countries/. Accessed 10 June 2024.

Mothiba, M., Moja, S., and Loans, C. (2017) A Review of the Working Conditions and Health Status of Waste Pickers at Some Landfill Sites in the City of Tshwane Metropolitan Municipality, South Africa, *Advances in Applied Science Research, 8 (3)*: 90–97.

National Minerals Information Center. (2018). *Asbestos Statistics and Information: Minerals Yearbook: Advance Data Release of the 2018 Annual Tables.* National Minerals Information Center.

Nixon, R. (2011). *Slow Violence and the Environmentalism of the Poor.* Harvard University Press.

Orphan Well Association. (2023). *Orphan Inventory.* www.orphanwell.ca/about/orphan-inventory. Accessed 13 June 2024.

Pasley, J. (2020). Inside Louisiana's Horrifying 'Cancer Alley', an 85-Mile Stretch of Pollution and Environmental Racism That's Now Dealing with Some of the Highest Coronavirus Death Rates in the Country, *Business Insider*, 9 April. www.businessinsider.com/louisiana-cancer-alley-photos-oil-refineries-chemicals-pollution-2019-11 Accessed 6 June 2024.

Reeves, H. (2001) A Trail of Refuse, *New York Times Magazine*, 18 February. http://large.stanford.edu/publications/coal/references/reeves1/. Accessed 27 October 2024.

Riofrancos, T., Kendall, A., Dayemo, K.K., Haugen, M., McDonald, K., Hassan, B., Slattery, M., and Lillehei, X. (2023). *Achieving Zero Emissions with More Mobility and Less Mining.* Climate Community Project. www.climateandcommunity.org/more-mobility-less-mining. Accessed 14 June 2024.

Sandlos, J. and Keeling, A. (2012). Claiming the New North: Development and Colonialism at the Pine Point, Mind, Northwest Territories, Canada, *Environment and History, 18 (1)*: 5–34.

Scheidel, A., Del Bene, D., Liu, J., Navas, G., Mingorría, S., Demaria, F., Avila, S. et al. (2020) Environmental Conflicts and Defenders: A Global Overview, *Global Environmental Change, 63*: 102104.

Veracini, L. (2010). *Settler Colonialism: A Theoretical Overview.* Palgrave MacMillan.

Veracini, L. (2011). Introducing Settler Colonial Studies, *Colonial Studies, 1 (1)*: 1–12.

Weisgall, J. (1994). *Operation Crossroads: The Atomic Tests at Bikini Atoll.* Naval Institute Press.

Wilkes, J. and Hird, M.J. (2019) Colonial Ideologies of Waste: Implications for Land and Life, *EuropeNowJournal*, 27 (May). www.europenowjournal.org/2019/05/06/confronting-waste/.

Williams, I. (2020). One Man's Trash is Another's Burden: Social Justice and Waste Management, *Popular Education*, 19 February. https://populationeducation.

org/one-mans-trash-is-anothers-burden-social-justice-waste-management/. Accessed 6 June 2024.

World Economic Forum. (2020). *What Is Environmental Racism and How Can We Fight It?* www.weforum.org/agenda/2020/07/what-is-environmental-racism-pollution-covid-systemic/. Accessed 4 June 2024.

Wynter, S. (n.d.). *Black Metamorphosis: New Natives in a New World.* Institute of the Black World Records, MG 502, Box 1. Schomburg Center for Research in Black Culture.

Yeung, J. and Seo, Y. (2024). North Korea Trash Balloons are Dumping 'Filth' on South Korea, *CNN*, 29 May. www.cnn.com/2024/05/29/asia/north-korea-trash-balloons-intl-hnk/index.html. Accessed 25 June 2024.

Yusoff, K. (2018). *A Billion Black Anthropocenes or None.* University of Minnesota Press.

6

REDUCING WASTE

OVERVIEW

Chapter 6 focuses on the Waste Hierarchy's apex: waste reduction. In order to live within our planet's carrying capacity, we must extract, produce, distribute, and consume less. The global waste crisis is endemic to the United Nations' Sustainable Development Goals (SDGs). This book has been mainly concerned with SDG 6: clean water and sanitation. The final chapter of this book focuses on SDG 12: responsible production and consumption; and SDG 13: climate action. Chapter 6 argues that consumption cannot meaningfully be separated from resource extraction, production, and transporting products for retail. Most people on earth are consuming within planetary limits, while a few million people are very significantly overconsuming. Consumers with the income and time resources may make purchasing decisions that are less detrimental to the environment. Beyond individual consumer behavior, people may also effectively contribute to community, non-governmental organizations, and other efforts to pressure governments and industry to produce fewer, and more eco-friendly, products. Using the examples of mobility (how we get people and things from place to place) and plastics bans, this chapter shows that individual action alone cannot resolve our waste crisis or effectively achieve climate action.

DOI: 10.4324/9781003398424-6

INTRODUCTION

In September 2015, the United Nations adopted seventeen Sustainable Development Goals (SDGs) consisting of:

1 No Poverty
2 Zero Hunger
3 Good Health and Wellbeing
4 Quality Education
5 Gender Equality
6 Clean Water and Sanitation
7 Affordable and Clean Energy
8 Decent Work and Economic Growth
9 Industry, Innovation and Infrastructure
10 Reduced Inequalities
11 Sustainable Cities and Communities
12 Responsible Production and Consumption
13 Climate Action
14 Life Below Water
15 Life on Land
16 Peace, Justice and Strong Institutions
17 Partnerships for the Goals

The UN publishes yearly updates on global progress made towards the SDGs. While SDG 6 – Clean Water and Sanitation – most obviously concerns waste issues, waste is implicated in all of the SDGs. Life Below Water (SDG 14) is profoundly impacted by the proliferation of plastics and microplastics in our oceans. The job of waste picking reveals the urgent need for Gender Equality (SDG 5), Good Health and Wellbeing (SDG 3), Decent Work and Economic Growth (SDG 8), and No Poverty (SDG 1).

Thus far, all of the chapters in this book have been directly devoted to SDG 6: Clean water and Sanitation. Understanding the basics about waste means getting to the context within which waste is produced and managed. Waste does not just happen. The current global waste crisis has been manufactured and structured to produce waste and pollution. This final chapter focuses on SDGs 12 and 13: Responsible Production and Consumption, and Climate Action, respectively.

SDG 12: RESPONSIBLE PRODUCTION AND CONSUMPTION

While all human beings consume in order to survive, some people consume more – and for some, much more – than they need to live. About twenty percent of the world's population consumes about eighty percent of the world's resources (Acconia n.d.). The richest one percent of people are creating more greenhouse gas emissions than the poorest fifty percent of people (United Nations n.d.). For decades, the non-profit organization Global Footprint Network has been comparing human consumption with the Earth's carrying capacity. In his book, *The Day the World Stops Shopping* (2021: 32–35), J.B. MacKinnon argues that if everyone lived like the average American, we would need five Earths in order to sustain us. We need four or more Earths to live like Sweden, South Korea, Australia, or Canada. We need more than three planets to live like people in Germany, the Netherlands, Russia, Finland, or Norway. We need about two planets to live like the average person in China. Put another way: each person on Earth can use up 1.6 global hectares worth of resources. Americans use up eight global hectares.

By contrast, people in some fifty-three countries (including Ecuador, Cuba, Sri Lanka, Armenia, the Dominican Republic, the Philippines, Jamaica, Indonesia, and Egypt) consume at or below the one-planet level, and account for almost half of the world's population. Of course, as Chapter 5 demonstrates, there are wide differences in wealth and consumption within any given country, such as the United States, Canada, the United Kingdom, and France.

Yet, consumption patterns do not positively correlate with people's happiness. There is evidence to suggest the opposite: most highly developed consuming countries do not make the top ten of the Happy Planet Index (n.d.). Developed by the Wellbeing Economy Alliance, the Index measures the wellbeing of a nation's people multiplied by their life expectancy and divided by their ecological footprint (Happyplanetindex.org). Among the top ten countries are (in descending order) Costa Rica, Vanuatu, Colombia, Switzerland, Ecuador, Panama, Jamaica, Guatemala, Honduras, and Uruguay. Out of 152 countries, the United States ranks near the bottom, at 122.

RESPONSIBLE CONSUMPTION

There is no doubt that consumption levels in affluent countries, and particularly amongst affluent people within affluent countries, has dramatically increased (see Chapter 2). Criticism is leveled at countries like China that are producing more, and more toxic waste. But this must be placed in the context that 'a large part of the toxic production in the South exists to meet the demand from rich countries... It is the rich who destroy the planet and externalize the damage' (Jarrige and Le Roux 2020: 309). As Chapter 3 examines, manufacturing and retail industries very successfully market consuming as key to individual choice, efficiency, and freedom. Indeed, consuming in particular is marketed as emblematic of freedom in Western countries. Now, an increasing number of individuals recognize the connections between consumption and environmental degradation, including the climate crisis. Consuming green, repair shops, and other initiatives tend to target consuming and consumerism at the individual level.

Individuals and households with the money, time, and other resources are making lifestyle choices (types of vehicles driven, amount of energy used in the home, length of showers, diet, and so on) that decrease their carbon footprint and are otherwise less harmful to the environment. In other words, for the most part, sustainable living in industrialized high-income countries typically means individuals altering their consumption habits. When these alterations lead to decreased consumption, then environmental benefits include decreases in energy use and waste, as well as the potential psychological benefits to individuals knowing that they are making a positive impact on the environment.

Yet, since capitalism is predicated on ever-increasing consumption, we anticipate that our global political-economic system seeks ways of enveloping attempts to decrease consumption. And, in fact, capitalism does exactly this by harnessing neoliberalism's focus on individual responsibility and entrepreneurship. For example, the Zero Waste Movement focuses on approaches such as product eco-design, using innovative technologies to produce products within a closed-loop supply chain, and ascertaining the environmental impact of any given product through Life Cycle Analysis (Song et al. 2015). The main objective of Zero Waste is to divert as much generated waste as possible. However, Zero Waste 'lifestyle influencers' and community organizations tend to focus more on 'green consuming' and recycling rather than consuming less.

As such, people are often ironically encouraged to consume *more* to achieve zero-waste lifestyles. Household goods such as metal and glass storage containers for bulk goods, reusable containers, brushes made from wood and natural fibers etc. are all recommended and otherwise marketed. As such, the Zero Waste movement has created a new marketing *niche*. Granted, the movement does advocate shopping at refilleries or other places that offer goods in bulk, thus reducing packaging, and shopping at thrift stores. But Zero Waste's major focus is to divert waste away from landfill, and not consuming less per se. As such, the Zero Waste movement favors consuming differently over consuming less. Simply re-classifying waste as recycling or energy does not require (or even encourage) decreased consumption but is rather a way of 'hiding behind' diversion rather than tackling the challenge of reducing the production of products. Zero Waste individual behaviors and/or group initiatives are largely dependent upon recycling, which comes with considerable environmental costs (see Chapter 4). Indeed, Samantha MacBride identifies recycling as 'busyness': a 'sense that progress is being made' even though the facts show that it is not (2012: 171).

Chapter 2 shows that even when individuals attempt to consume responsibly, they are often forced to consume items due to the planned obsolescence of those purchased items. A well-made item will last a lifetime and only be purchased once – great for the consumer, not so great for the manufacturer and capitalist profit (it is more profitable for some people to buy the same product repeatedly than for more people to buy a product only once). Thus, as well as sophisticated marketing strategies, and the constant introduction of 'new' products (since its first version served in 1886, Coca-Cola has created over 100 different variations of this single beverage product), manufacturers rely on planned obsolescence to increase consumption.

Like attempts to reduce individual and household consumption, people are increasingly pushing back against planned obsolescence. For instance, in 2018 the Italian Competition Authority fined both Apple and Samsung for unfairly promoting software updates that seriously impaired the functioning of customer cellular devices (Busch et al. 2018). In an attempt to reduce e-waste, the European

Union passed an amendment in 2022 that all cellular devices released after 2024 have to be equipped with universal charging ports (Fourneaux 2024). Progress is also being made to provide consumers with a greater right to repair goods. In 2023, the government of France earmarked 154 million euros in a five-year fund to financially incentivize people to repair their clothes and shoes (Willsher 2023). This initiative aims to provide a repair bonus (people will be able to claim back between six and twenty-five euros for the cost of repairs), support workers and retailers who offer repairs, create new jobs, reduce waste in the fashion industry, and promote a more Circular Economy. In the same month, Canada's province of Québec passed Bill 29: An Act to Protect Consumers from Planned Obsolescence and to Promote the Durability, Repairability and Maintenance of Goods. The increasing appearance of Right to Repair legislation in more countries means that consumers will have greater control over their consumption of goods, and more services will be available to them.

As with Right to Repair initiatives, the Sharing Economy aims to reduce consumption. Many large cities already have bicycle share programs, in which consumers pay a fee to rent bicycles that are distributed throughout the city, thereby reducing the number of bicycles that are privately owned (Murphy and Usher 2015). Car-sharing in many European countries is successfully increasing the use of shared electric vehicles, reducing the mobility cost to users, reducing the number of privately owned cars, and reducing greenhouse gas emissions (Roblek et al. 2021). The Sharing Economy is also being applied to food service ware. Canada's capital city of Ottawa is piloting a first-of-its-kind collaborative program with major grocers. This program gives consumers the opportunity to purchase selected products in free reusable containers, which they can then return to 'smart' tracking bins for collection, sanitization and redistribution by the grocery service partner. If successful, the plan is to expand the system to include more retailers (such as restaurants that offer meals to go) as well as expand over a greater geographic area (Circular Innovation Council 2024). On a smaller though similar scale, the tourist city of Banff, Alberta, Canada is currently piloting a reusable cup program to reduce the number of disposable coffee cups thrown out on a daily basis (Rhode 2024).

Individual and household behavior focused on reducing consumption is beneficial to the environment. We also need to understand these actions in their broader context. For example, at least one study found that bicycle share users are mostly middle- and upper-class males, suggesting inequity in the ability of consumers to access the program (Murphy and Usher 2015). Individuals and households who are able to make choices to consume less and/ or consume in ways that are less detrimental to the environment are more likely to be people who have the financial means and resources to do so. At least one parent must have sufficient leisure time to shop 'green'. According to the US Department of Agriculture Economic Research Service (2009), about 2.3 million people (just over two percent of all households in the United States) live in 'food deserts': these millions of people lack access to healthy food options such as fruits and vegetables, and affordable food. What people do have access to tends to be more take-out fast-food that is more highly packaged, creating more waste.

And though reuse programs, such as cup-sharing and returnable food containers, do reduce the use of resources needed to produce these products, the tradeoff is a greater use of water and energy (for washing) and transportation of reusable containers from the return bins to the sanitation facility and back to the distribution centers and grocery stores. If the sanitation centers are not local, significant carbon emission costs can be incurred. It is crucial, when setting such programs up, that all environmental costs are evaluated to ensure that overall consumption is truly being reduced rather than simply projecting an appearance that we are consuming less.

Mobility is a good illustration of these arguments. The transition to electric vehicles is a good way of illustrating that individuals consume (and therefore waste) within a broader context. This context is neo-liberal capitalism. And this context subscribes to the simplistic – and economics-driven – view that overconsumption and other problems can be resolved through modifying individual behavior, and that this modification can be effectively achieved through economic incentives (e.g. variable tax rates; see, for instance Stern 2007).

The term 'excess capitalism' refers to the late twentieth century's creation of leisure and other activities that created new consumption markets. Before the creation of leisure time, people (including children old enough to labor) worked either every day of the week, or reserved one day per week to attend religious services. As such, any leisure activities (county fairs for instance) took

place infrequently and locally within walking or bicycling dis-
tance. As John Urry (2010) explains, towards the end of the last
century, wealthy countries such as the United Kingdom designed
and developed seaside resorts in places like Blackpool specifically
for working class mass leisure, and were premised on the contrasts
between work and holiday, home and away, ordinary time and leisure time, and workspace and holiday space. Working class holidays
are 'disciplined pleasure with conspicuous consumption specific to
individuals' (ibid: 201). Excess capitalism is achieved, in part, through
the ascension of neoliberalism, which always seeks to de-regulate
and remove state or collective services, control, and oversight. Neoliberalism leaves the private sector to assume this control as the best
and most 'natural' way for societies to operate. As David Harvey puts
it, neoliberalism is now 'incorporated into the common-sense way
many of us interpret, live in, and understand the world' (2005: 3).
Neoliberalism has generated consumption excess destinations:

> There are distinct zones that are travelled to, often made possible
> by large infrastructural projects; there are gates controlling entry
> and exit; they are highly commercialized with many simulated
> environments; there is only pleasure, no guilt; norms of behaviour
> are un-regulated by family/neighbourhood; there are liminal
> modes of consumption; bodies are subject to commodifica-
> tion; there is digital control; these sites are increasingly globally
> known for their consumption/pleasure excess; and they are sites
> of potential mass addiction.
>
> (Urry 2010: 203)

More people traveling by private vehicles means more consumption and more waste. More people become occupied with acquiring
products and services beyond the local neighborhood or community in relation to other people doing the same. Mobility has
greatly expanded from local to regional to national to international
destinations. Not only are people with the means travelling further
distances but consuming more as part of the leisure experience. The
working-class getaway within any given country (Blackpool, Las
Vegas, Niagara Falls) has expanded to 'sunshine destinations' such as
Majorca, Tenerife, and Lanzarote. Those with the money can visit
Dubai, which explicitly describes itself as the world's top excessive

consumption destination complete with human-constructed islands, a seven-star hotel, a domed ski resort, the world's tallest building, and much more (see Facebook's Do Buy: Dubai and YouTube's Do Buy! A Dubai Documentary). As Mike Davis succinctly observes, Dubai:

> is the apotheosis of the neoliberal values of contemporary capitalism... Dubai, indeed, has achieved what American reactionaries only dream of – an oasis of free enterprise without income taxes, trade unions, or opposition parties (there are no elections). As befits a paradise of consumption, its unofficial national holiday, as well as its global logo, is the celebrated Shopping Festival, a month-long extravaganza sponsored by the city's twenty-five malls that begins on January 12 and attracts four million up-scale shoppers, primarily from the Middle East and South Asia.
>
> (2007: 60)

All of this mobility and product and service consumption requires more energy, and produces more waste. We may add cruise line holidays and 'all-inclusive' resorts to this increasing mobility and excess consumption. The Royal Caribbean Group's Royal Caribbean International Icon of the Seas cruise ship, for instance, can fit 7,600 passengers and has 2,805 staterooms (Royal Caribbean International n.d.). And in what's known as 'casino capitalism', developers constantly seek to emulate exclusive high-end destinations with mass-market versions of consumption excess.

We now see an urgent push to move away from fossil fuels through the transition to electronic vehicles. As Chapter 1 details, the fossil fuel industry is responsible for the lion's share of global climate change, producing over seventy-five percent of greenhouse gas emissions and almost ninety percent of all carbon dioxide emissions (United Nations n.d.). Fossil fuels are used to generate power (electricity and heat) by burning coal, oil, and gas. Residential and commercial buildings alone consume over half of all electricity. Most products are manufactured by machines that use fossil-fuel energy. The plastics industry uses chemicals sourced from fossil fuels, making manufacturing one of the largest sources of greenhouse gas emissions worldwide. Global deforestation and agriculture – mainly achieved using fossil-fuel powered machines, account for more than a quarter of the remaining greenhouse gas emissions.

As well, the fossil fuel industry subtends global land, sea, and air transportation, as most cars, trucks, planes, and ships derive their energy from fossil fuels. Transportation alone accounts for almost a quarter of all carbon dioxide emissions. And much of this transportation is in the form of private vehicles. The push to transition away from fossil-fuel-powered cars is leading to a surge in demand for electric vehicles. Electric vehicles are powered with batteries that use various chemicals to share electrons, including lithium phosphate, lithium manganese oxide, lithium titanate, nickel manganese cobalt, and nickel cobalt aluminum (US Department of Energy n.d.). The United States, the European Union, China, the United Kingdom, Canada, and other countries increasingly define these as 'critical minerals' to signify their economic importance and vulnerability to supply chain disruptions and national security (Riofrancos et al. 2023). As wealthy nations scramble to secure sufficient minerals needed for electric vehicles, they are seeking to dominate regions of the world where these minerals – and lithium in particular – are sourced. Most lithium is found in Australia, Chile, China, Argentina, Brazil, Zimbabwe, the United States, and Portugal, with other countries such as Canada making strong efforts in lithium site exploration. Chapter 5 explores the human rights violations and environmental costs of mining, and especially the production of vast quantities of waste.

If we are not to increase these human and environmental costs, and if we are to meet the demand for electric vehicles, then the transition cannot simply be from fossil-fuel-powered vehicles to lithium-powered vehicles. In their detailed study of decarbonized mobility scenarios in the United States, Thea Riofrancos et al. (2023) found that transitioning to electric vehicles while remaining dependent on private transportation will not achieve either net-zero or significantly mitigate climate change. As well, as long as these private vehicles use large batteries (77 to 123 kWh), the demand for lithium will triple our current demand, and bring with it increased regional and global political instability, as well as increased mining and transportation waste. That is, in order to meet the SDG 1 commitment to No Poverty, SDG 7 to Affordable and Clean Energy, SDG 11 to Sustainable Cities and Communities, SDG 12 to Responsible Production and Consumption, SDG 13 to Climate Action, and SDG 16 to Peace, Justice, and Strong Institutions, individuals need to switch to mass public transportation, cycling, and walking. Governments need to do vastly more to provide active and public transport, making private vehicle ownership more expensive and far less convenient. In terms of transportation, then, responsible consumption does not mean buying an electric car (especially one with a medium to large battery): it means transitioning to public transit, bicycles, and walking whenever possible.

The waste of these places and spaces of excess is literal, in the form of infrastructure development (industrial, commercial, and institutional – ICI), food waste, products and services waste, as well as the waste from the vast energy consumption required to power all of this excess. But wasting is also an *explicit* part of the excess, in the form of having enough wealth to be able to holiday (i.e. waste time) and buy far more products and services than are needed. Conspicuous wealth must, by definition, be wasteful (Veblen 1912). As Urry notes:

> So the scale and impact of 'waste' production has moved dramatically upwards, especially so with this economy of waste focusing upon gambles on the future, whether on the casino table or global commodities or junk bonds or sub-prime mortgages. It is a kind of casual production and consumption as places come and go, being produced and then used up... One consequence of such systemic production and resulting high status from the consumption of excess and waste is to generate further upward shifts in carbon emissions.
>
> (2010: 206–207)

Neoliberal capitalism itself markets excess capitalism as freedom. As Joseph Stigliz argues:

> Understanding the meaning of freedom is central to creating an economic and political system that delivers not only on efficiency, equity, and sustainability but also on moral values. Freedom – understood as having inherent ties to notions of equity, justice, and well-being – is itself a central value. And it is this broad notion of freedom that has been given short shrift by powerful strands in modern economic thinking – notably the one that goes by the shorthand term *neoliberalism*, the belief that the freedom that matters most, and from which other freedoms indeed flow, is the freedom of unregulated, unfettered markets.
>
> (Stigliz 2024)

Economists, industry, and conservative politicians have touted the commitment to 'unfettered markets' – the idea that society works best when economic markets manage themselves through competition rather than being regulated by government. Market competition, according to neoliberalism, will produce the greatest efficiency and the greatest individual freedom (Freidman 1962; see also Stedman Jones 2014). Individual freedom here emphasizes consumer choice: from a dizzying array of breakfast cereals to holiday destinations. The 'freedom' to buy everything we want, when we want, and even the freedom to incur debt in the pursuit of consumption. But this, as Stigliz (2024) observes, 'has led to the freedom of a few at the expense of the many', or as Isaiah Berlin famously wrote, 'freedom for the wolves has often meant death to the sheep'. And so, wealthy individuals, communities, and nations practice what Jennifer Clapp (2002) refers to as 'waste distancing' whereby waste is exported away from its sources and on to people and environments who lack the power to reject the waste and its pollution.

More generally, consumers are charged with complex and difficult trade-offs, while manufacturers are largely protected by the capitalist principle that any and all technological inventions that are profitable are therefore good. Thus, industries market consumption as a choice between buying a fossil-fuel- or electric-powered vehicle. The choice to pay more for their airline ticket in order to offset the carbon emissions of their travel. The bigger questions about adequacy-oriented living instead of excess-oriented living are left purposefully unasked (Schulz, Hjaltadóttir, and Hild 2019).

RESPONSIBLE PRODUCTION

Compared with consumption, responsible manufacturing receives far less media and public attention. As Nicolas Graham points out, 'The cultural criticism of consumerism is important, but without an analysis of the deep complementarity of production-consumption we overdetermine consumption' (2021: 106). In other words, there is a stark unevenness to the focus on individuals and households to shoulder the responsibility for pollution and waste and the environment more generally. As Katharine Owens and Katie Conlon put it, 'If you walked into your bathroom to find your tub overflowing with water, would you first begin mopping up the water on the ground, or would you turn off the tap?' (2021: 5). The global waste crisis will not be resolved until we turn off the waste-generating tap. And to do this, we need to consider what motivates producers to create products with known and unknown contaminants, and waste more generally.

The short answer to this question is capitalism. As this book demonstrates, capitalism is not concerned with waste beyond profit. That is, capitalism is concerned with realizing any potential that waste may offer to increase capital. Thus, there is a growing interest in waste-as-resource (see Chapter 3) because this economizes waste. Companies derive profit from transforming waste into energy (electricity for instance), and in transporting waste away from some regions (wealthy) to other regions (poor). This is where capitalism's interest in waste begins and ends. Capitalism does not concern itself with the environmental or human health consequences of either production or consumption. Companies almost exclusively adhere to this focus on profits at the expense of the environment and human health.

The public has known about global climate change since the first Intergovernmental Panel on Climate Change (IPCC) Report in 1990. That was thirty-four years ago. Climate change is inseparable from the global energy crisis because fossil fuel production and consumption are responsible for the lion's share of carbon emissions (Homer-Dixon 2006). Fossil fuels power virtually all of our mobilities (air, sea, and land travel – see previous section), resource extraction, manufacturing, agriculture, health care, and so on. Climate change scientific research is demonstrating that as well as negative feedback loops which restore equilibrium, positive feedback loops are increasing temperatures: as George Monbiot (2007) states, climate changes begets climate change. And political conservatives undermine climate change information (McCright and Dunlap 2010).

The fossil fuel industry is disproportionately responsible for carbon emissions and waste generation. Research demonstrates that fossil fuel companies have known about their contribution to climate change for over fifty years. In 2017, Geoffrey Supran and Naomi Oreskes published an in-depth study of ExxonMobil Corp's climate change communications, which the researchers updated in 2020. In their exhaustive analysis of internal communications within this major fossil fuel company (formerly Exxon and Mobil corporations), the researchers found that the company was very well aware – since at least the 1970s – that fossil fuels are major contributors to climate change. At the same time, this company knowingly misled the public about their contribution to climate change by financing public communications that cast doubt on the reality of climate change. The researchers empirically analyzed 1448 advertorials (paid editorial-style advertisements in the *New York Times*, *Washington Post*, and other newspapers) that attempted to undermine scientific data on anthropogenic global warming. They also analyzed thirty-two internal memos, seventy-two peer-reviewed articles, and forty-seven non-peer-reviewed articles, as well as company reports, webpages, and company speeches. Of the peer-reviewed documents, eighty-three percent acknowledge anthropogenic global warming. Of the non-peer-reviewed documents, the corporation acknowledges it in sixty-six percent, seventeen percent 'acknowledge and doubt', and seventeen percent doubt. This difference can be accounted for by the fact that peer-reviewed articles are reviewed by experts, whereas non-peer-reviewed articles are not. Peer-review makes it much more challenging for companies to deny scientific facts. Finally, eighty percent of Exxon's internal documents acknowledge their contribution to climate change, while fifteen percent express 'reasonable doubt' (Supran and Oreskes 2020: 6).

In contrast, seventy-two percent of the advertorials cast doubt on anthropogenic global warming, while five percent doubted and acknowledged and twenty-one percent acknowledged. Doubt took several forms, including, for instance, calling climate change modeling 'unreliable' and claiming that scientists are 'not yet capable of predicting Earth's global climate' (ExxonMobil 2000a, 2000b). The advertorials pushed a rhetoric of climate change risk rather than certainty: akin to arguing that death is a risk. Part of the risk rhetoric is to pit climate change reduction against poverty reduction and economic growth. They also attempt to cast doubt on the solvability of climate change, claiming that carbon reduction goals are either impossible or unrealistic.

ExxonMobil Corp swiftly attempted to undermine these researchers' methods and findings. For instance, the corporation argued that the researchers 'obscure[ed] the separateness of the two corporations', which invalidated their conclusions (Bacckelmans 2019; Swarup 2020; see Mulvey 2017). The researchers point out

that when Exxon and Mobil merged, the ExxonMobil Corporation inherited the legal responsibility for the companies: indeed
the research clearly demonstrates that ExxonMobil Corp continued the misinformation campaign after the merger. It is also clear
from the internal documents that Exxon, Mobil, and ExxonMobil
Corp had accessed and understood the mainstream scientific data
and knowledge about climate change (Supran and Oreskes 2020;
Supran, Rahmstorf, and Oreskes 2023). Indeed, the companies have
actually funded scientific research demonstrating climate change,
while also funding groups and individuals and participating in organizations that deny climate change.

Research clearly shows that fossil fuel companies worldwide,
and including ExxonMobil Corp, are increasing – and plan to keep
increasing – fossil fuel production. According to the 2023 Production Gap Report (SEI et al. 2023), governments with fossil-fuel producing regions plan to *double* fossil fuel production by 2030. Global
coal production will increase until 2030, and oil and gas production until 2050. These same governments are pledging 'net-zero'
emissions through carbon capture and storage and carbon dioxide
removal; technologies that are proven ineffective in reducing carbon
emissions. This means that, despite the Paris Agreement's pledge to
keep Earth's temperature to a 1.5 degree Celsius increase, we are
way off this target.

SDG 13: CLIMATE ACTION

As Chapter 1 details, waste is critical to climate action. Waste includes
all of the materials that are discarded in the processes of resource
extraction, manufacturing, retailing, and post-consumption. Greenhouse gas emissions are also a form of waste. According to the US
Environmental Protection Agency (2024), up to half of global greenhouse gas emissions arise from resource extraction, manufacturing,
fuels, and food. Reducing waste reduces greenhouse gas emissions:
waste reduction is integral to climate action.

Waste reduction and its critical climate change benefits will not
result from individual and household behavioral change alone. Nor
can we rely on companies to voluntarily reduce the waste they produce, or take meaningful climate action. In a study of 1,925 large-cap
firms, Yu et al. (2020) found that a lack of standardization in

disclosure rules of sustainability data, and no global governing body to ensure accuracy of reporting by companies about their efforts to be more environmentally conscious, resulted in more than half of the companies engaging in greenwashing. The fossil fuel/plastics industry is a primary example.

Facing increasing pressure from environmental groups such as Greenpeace, Friends of the Earth, Plastic Pollution Coalition, Plastic Change, and the World Wildlife Fund, governments, and the public, the contemporary plastics industry is turning their attention to addressing public concerns about plastics waste. In 2018, Canada led the way as G7 President in developing the Open Plastics Charter, establishing a framework for sustainable plastics use and eliminating plastics litter on land and at sea. Globally, twenty-eight govern-ments and over seventy companies/organizations have endorsed the Charter. In March 2022, the UN Environment Assembly adopted a resolution to combat plastics pollution with a global and legally binding plastics treaty by 2024 that will take into consideration the whole plastic life cycle with internationally binding targets (Berg-mann et al. 2022; UNEA UNEP 2022).

However, the Center for International Environmental Law (2024) found the following concerning information: at the April 2024 Fourth Session (INC-4) of the UN's Environment Programme meetings in Ottawa, Canada, whose purpose was to develop an international and legally binding document on plastics pollution, there were 196 lobbyists from the fossil fuel and chemical industries. This number of lobbyists is three times greater than the fifty-eight independent scientists from the Scientists' Coalition for an Effec-tive Plastic Treaty, and seven times greater than the twenty-eight representatives of the Indigenous Peoples Caucus. The fossil fuel and chemical industries had more lobbyists at the INC-4 meetings than the smallest eighty-seven country delegations *combined*. And the fossil fuel and chemical industry lobbying was successful: after a full week of meetings, the INC-4 failed to include any provisions for the reduction of plastics or polymer production.

This is one example of a long-standing pattern whereby com-panies market themselves as environmentally conscious on social media, while at the same time actively resisting taking responsibility for the environmental costs of their products' life cycles. The plas-tics industry is a sort of subdivision of the larger fossil fuel industry.

A growing chorus of scholars and environmental groups are advocating that brand names (Coca-Cola and so on) take financial and environmental responsibility for the environmental pollution their products create (see for example BreakFreeFromPlastic 2023; Cowger et al. 2024; Charles, Kimman, and Saran 2021). Citizens need to join this chorus in demanding that the governments that represent them stop providing regulator concessions, and instead hold polluting companies accountable for their waste production.

INDIVIDUAL AND SYSTEM CHANGE

George Monbiot argues that what separates humans as a species is our failure to prioritize our own species survival (2021) by allowing ourselves to be manipulated by industry into believing that individual changes in our consumer behavior will make a real difference to the ecological health of our planet: 'tiny issues such as plastic straws and coffee cups, rather than the huge structural forces driving us towards catastrophe'. He goes on to argue that our focus on 'microscopic solutions' is not accidental:

> All of us are expert at using the good things we do to blot out the bad things. Rich people can persuade themselves they've gone green because they recycle, while forgetting that they have a second home (arguably the most extravagant of all their assaults on the living world, as another house has to be built to accommodate the family they've displaced). And I suspect that, in some deep, unlit recess of the mind, we assure ourselves that if our solutions are so small, the problem can't be so big... The great political transition of the past 50 years, driven by corporate marketing, has been a shift from addressing our problems collectively to addressing them individually. In other words, it has turned us from citizens into consumers. It's not hard to see why we have been herded down this path. As citizens, joining together to demand political change, we are powerful. As consumers, we are almost powerless.
>
> (Monbiot 2021)

Capitalism will always advance waste management regulations, policies, and practices that benefit capitalism. Reduction does not benefit capitalism. Therefore, individuals and communities – the public – need to think about what regulatory and policy incentives (carrots) and penalties (sticks) to adopt in order to hold industry responsible for the products they produce, and all of the marketing that encourages, incites, shames, flatters, and otherwise increases consumption of their products. We need national and international regulations that require producers to build product waste and contamination into their product production projections. All of these amount to one overarching goal: to get resource extraction companies and manufacturers to take responsibility for the waste that they produce.

One of the major means of achieving this goal is public pressure. When enough people voice dissatisfaction or make a demand, democratic governments respond to their voters. For instance, extended producer responsibility (EPR) regulations are a response to public pressure. EPR shifts responsibility for any given product's life-cycle on to the companies that manufacture these products. In 2022, the European Union proposed the EU Packaging Regulation that would make all packaging recyclable by 2030, restrict unnecessary packaging, and promote reusable and refillable packaging solutions. The Regulation would also create mandatory rates of recycled content that producers have to include in new plastics packaging, with the target being to reduce packaging waste by fifteen percent per capita per member state by 2040 (European Commission 2022). The EU's Waste from Electrical and Electronic Equipment (WEEE) Directive and EU Battery Directive focus on specific products and their waste. In response, some companies are investigating ways to eco-design product packaging and are otherwise moving towards less packaging, and refill and return options.

DEGROWTH AND THE GREEN NEW DEAL

One of the major ideas for merging individual and producer responsibility together with government support is the Degrowth movement. Degrowth begins with the acceptance that capitalist economic growth causes pollution, and that pollution is unevenly distributed globally (Hird 2022; Liboiron 2021; Nixon 2011). As Samantha MacBride points out 'industrial zones – the only suitable spots for large-scale processing of recycling as well as garbage transfer, disposal, and incineration – are overwhelmingly near the homes of people of colour and sometimes working-class white people' (2012: 125).

Rejecting the individual consumer behavioral changes and techno-fix solutions that the current neoliberal capitalist system relentlessly repeats, degrowth confronts the system itself. Degrowth recognizes that our global societal metabolism exceeds Earth's carrying capacity (Schmelzer, Veller, and Vansintsan 2022). Rather than echoing industry and capitalist governments' emphasis on 'green consumer' responsibility, degrowth emphasizes waste producer responsibility.

Degrowth takes several forms, which may be generally understood as 'bottom up' or 'top down' approaches. Bottom up approaches include initiatives such as Slow Shopping (as part of the Slow Life movement) that encourages people to take more time while shopping and to reflect more seriously about whether they actually need a particular product (Fulenwider 2016). Low Tech Magazine and the International Centre for Anti-Consumption Research are fueling ways of decreasing extraction, production, consumption, and waste. These bottom up initiatives have the advantage of flexibility and adaptability to local contexts. Their disadvantage is that they tend to be small movements, and rely on volunteers (who tend to be middle-class women with the financial means to volunteer their skills). Critics also point out that Voluntary Simplicity, Slow Shopping, and other movements apply to about seventeen percent of the world's population: the other eighty-three percent are already practicing involuntary simplicity (MacKinnon 2021).

Degrowth at a much larger scale takes the form of the Green New Deal (GND). The GND focuses on clean energy, green infrastructure, and an economically, politically, and socially equitable society. Supporting, augmenting, and creating green infrastructure (free mass transit for instance) creates 'green collar jobs' (Durning 1999). In terms of waste, degrowth 'is based on the daily labour of reduction and reuse… it envisions a society that reuses materials to satiate needs' (Savini 2023: 2). Reduction would have a profound impact on waste generation: for instance, '[s]hutting down worldwide clothing production for a year would be equal to grounding all international flights and stopping all maritime shipping for the same time period' (MacKinnon 2021: 157). Controlling the waste management industry requires supra- and national organizations and governments in cooperating to regulate – and enforce – such things as waste exports, and to shift technological innovation away from recycling and disposal and towards self-sufficient modes of sustainable urban and rural living (Hird and Dee 2024).

REVIEW

People with the means to make choices about what they consume may consume less. These same people have reported that they are willing to pay more for environmentally friendly products and may be more likely to purchase products from firms they consider to be socially responsible (de Freitas Netto et al. 2020). Some people may also have the means to consume differently, consuming products whose life cycle has a less negative environmental footprint than other products. People with the means – the upper and middle classes of any given country – are also more likely to draw the attention of their government representatives, who court their vote at each election cycle. This means that people who are concerned about the environmental impacts of producing and consuming products may join their voices and signatures to environmental initiatives that call for greater industry and government responsibility in eliminating and better managing pollution and human health concerns. By creating binding regulations, as is being done in the European Union, consumers can have greater confidence that products are actually more environmentally sound and produced more sustainably than alternative products. That is, we need to take action to reduce consumption whenever we can, try to consume differently within our means, and join forces with others to advocate for a less contaminated and more socially just planetary system.

SUGGESTED READING

Klein, N. (2014). *This Changes Everything: Capitalism vs. the Climate*. Knopf Canada.

Magdoff, F. and Williams, C. (2017). *Creating an Ecological Society: Toward a Revolutionary Transformation*. Grassroots Books.

Narain, S. and Singh Sambyal, S. (eds.) (2016). *Not in My Backyard: Solid Waste Management in Indian Cities*. Centre for Science and Environment, New Delhi.

Rizzo, J. (2020). *Waste: Capitalism and the Dissolution of the Human in Twentieth-Century Theater*. Punctum Books.

REFERENCES

Acconia. (n.d.) *Natural Resources Deficit*. www.activesustainability.com/environment/natural-resources-deficit/?_adin=02021864894#. Accessed 14 March 2024.

Bacckelmans, N. (2019). *Letter from Nikolaas Bacckelmans (ExxonMobil, Vice President European Union Affairs) to Adina-Ioana Valean, MEP (Chair, Committee on the Environment, Public Health and Food Safety) and Cacelia Wikstrom, MEP (Chair, Committee on Petitions) of European Parliament*. https://perma.cc/VG6W-VFA5. Accessed 25 July 2024.

Bergmann, M., Almroth, B.C., Brander, S.M., Dey, T., Green, D.S., Gundogdu, S., Krieger, A., Wagner, M., and Walker, T.R. (2022). 'A Global Plastic Treaty Must Cap Production'. *Science, 376 (6592)*: 469–470.

BreakFreeFromPlastic. (2023). *Branded 6: Brand Audit Report*. BreakFreeFromPlastic.

Busch, C., De Franceschi, A., Durovic, M., Zuzak, J., Mak, V., Morais Carvalho, J., Nemeth, K., Podszun, R., and Riefa, C. (2018) 'Planned Obsolescence Challenging the Effectiveness of Consumer Law and the Achievement of a Sustainable Economy: The Apple and Samsung Cases'. *Journal of European Consumer and Market Law*, 7: 217–264.

Center for International Environmental Law. (2024). *Fossil Fuel Lobbyists Outnumber National Delegations, Scientists, and Indigenous Peoples at Plastics Treaty Negotiations*. www.ciel.org/news/fossil-fuel-and-chemical-industry-influence-inc4/. Accessed 19 October 2024.

Charles, D., Kimman, L., and Saran, N. (2021). *The Plastic Waste Makers Index: Revealing the Source of the Single-use Plastics Crisis*. Minderoo Foundation.

Circular Innovation Council. (2024). *Circular Innovation Council Collaborates with Canada's Biggest Grocers to Design an Innovative Reuse Program*. www.globenewswire.com/news-release/2024/01/17/2811057/0/en/Circular-Innovation-Council-Collaborates-with-Canada-s-Biggest-Grocers-to-Design-an-Innovative-Reuse-Program.html. Accessed 30 April 2024.

Clapp, J. (2002). The Distancing of Waste: Overconsumption in a Global Economy, in Princen, T. Maniates, M. and Conca, K. (eds.), *Confronting Consumption*. MIT Press.

Cowger, W. et al. (2024). 'Global Producer Responsibility for Plastic Pollution', *Sciences Advances, 10*: 1–7.

Davis, M. (2007). Sand, Fear, and Money in Dubai, in Davis, M. and Monk, D.M. (eds.), *Evil Paradises: Dreamworlds of Neoliberalism*. The New Press.

de Freitas Netto, S.V., Sobral, M.F.F., Ribeiro, A.R.B., and da Luz Soares, G.R. (2020). Concepts and Forms of Greenwashing: A Systematic Review. *Environmental Sciences Europe, 32*, 19. https://doi.org/10.1186/s12302-020-0300-3.

Durning, A.T. (1999). *Green-Collar Jobs: Working in the New Northwest*. Northwest Environment Watch.European Commission. (2022). European Green Deal: Putting an End to Wasteful Packaging, Boosting Reuse and Recycling, Press release, 30 November. https://ec.europa.eu/commission/presscorner/detail/en/ip_22_7155. Accessed 7 May 2024.

ExxonMobil. (2000a) Political Cart Before a Scientific Horse, *The New York Times*. Advertorial.

ExxonMobil. (2000b) Political Cart Before a Scientific Horse, *The Washington Post*. Advertorial.

Freidman, M. (1962). *Capitalism and Freedom*. University of Chicago Press.

Fourneaux, A. (2024). The European Union's New USB-C Standardization Amendment: What Does It Mean for Innovation Within the Consumer Technology Industry? *Northwestern Journal of International Law and Business*, *44 (1)*: 149–171.

Fulenwider, M. (2016). The Rising Influence of 'Slow Shopping Theory', *Business Today*. https://journal.businesstoday.org/bt-online/2017/the-rising-influence-of-the-slow-shopping-theory. Accessed 8 August 2023.

Graham, N. (2021). *Forces of Production, Climate Change and Canadian Fossil Capitalism*. Haymarket Books.

Happy Planet Index. (n.d.). *Happy Planet Index*. https://happyplanetindex.org/wp-content/themes/hpi/public/downloads/happy-planet-index-briefing-paper.pdf. Accessed 25 June 2024.

Harvey, D. (2005) *A Brief History of Neo-Liberalism*. Oxford University Press.

Hird, M.J. (2022). *A Public Sociology of Waste*. Bristol University Press.

Hird, M.J. and Dee, G. (2024). Mother Earth and Her Three Little Wasteful Pigs: Waste Reduction through Degrowth, in H. Corvellec (ed.), *Waste as Critique*. Oxford University Press.

Homer-Dixon, T. (2006). *The Upside of Down*. Routledge.

Jarrige, F. and Le Roux, T. (2020). *The Contamination of the Earth: A History of Pollutions in the Industrial Age*. The MIT Press.

Liboiron, M. (2021). *Pollution Is Colonialism*. Duke University Press.

MacBride, S. (2012). *Recycling Reconsidered*. MIT Press.

MacKinnon, J.B. (2021). *The Day the World Stops Shopping*. Vintage Canada.

McCright, A. and Dunlap, R. (2010). Anti-Reflexivity: The American Conservative Movement's Success in Undermining Climate Change Science and Policy, *Theory, Culture and Society*, *27 (2/3)*: 100.

Monbiot, G. (2021). Capitalism Is Killing the Planet – It's Time to Stop Buying into Our Own Destruction, *The Guardian*, 30 October. www.theguardian.com/environment/2021/oct/30/capitalism-is-killing-the-planet-its-time-to-stop-buying-into-our-own-destruction. Accessed 25 July 2024.

Monbiot, G. (2007). *Heat: How to Stop the Planet Burning*. Allen Lane.

Mulvey, K. (2017). ExxonMobil Attacks New Study that Exposes its Climate Deception…Again, *Union of Concerned Scientists*, 24 August. https://blog.

ucsusa.org/kathy-mulvey/exxonmobil-attacks-new-study-that-exposes-its
-climate-deceptionagain/. Accessed 25 July 2024.

Murphy, E. and Usher, J. (2015). The Role of Bicycle-Sharing in the City: Analysis of the Irish Experience. *International Journal of Sustainable Transportation, 9 (2)*: 116–125.

Nixon, R. (2011). *Slow Violence and the Environmentalism of the Poor.* Harvard University Press.

Owens, K. and Conlon, K. (2021). Mopping Up or Turning Off the Tap? Environmental Injustice and the Ethics of Plastic Pollution, *Frontiers in Marine Science, 8*: 1–8.

Rhode, M. (2024). 'Banff Borrows': How the Town Is Comparing Single-Use Waste. *Calgary Herald*, 19 April. https://calgaryherald.com/news/local-news/banff-borrows-how-the-town-is-combatting-single-use-waste. Accessed 30 April 2024.

Riofrancos, T., Kendall, A., Dayemo, K.K., Haugen, M., McDonald, K., Hassan, B., Slattery, M., and Lillehei, X. (2023). *Achieving Zero Emissions with More Mobility and Less Mining.* Climate Community Project. www.climateand-community.org/more-mobility-less-mining. Accessed 14 June 2024.

Roblek, V., Meško, M., and Podbregar, I. (2021) Impact of Car Sharing on Urban Sustainability. *Sustainability, 13 (2)*: 905.

Royal Caribbean International. (n.d.). *Ship Fact Sheets.* www.royalcaribbean-presscenter.com/fact-sheet/35/icon-of-the-seas/. Accessed 24 July 2024.

Savini, F. (2023). Futures on the Social Metabolism: Degrowth, Circular Economy and the Value of Waste. *Futures, 150*: 103018.

Schmelzer, M., Veller, A., and Vansintsan, A. (2022). *The Future Is Degrowth: A Guide to a World Beyond Capitalism.* Penguin.

Schulz, C., Hjaltadóttir, R.E., and Hild, P. (2019). Practising Circles: Studying Institutional Change and Circular Economy Practices. *Journal of Cleaner Production, 237*: 117749.

SEI, Climate Analytics, E3G, IISD, and UNEP. (2023). *The Production Gap: Phasing Down or Phasing Up? Top Fossil Fuel Producers Plan Even More Extraction Despite Climate Promises.* Stockholm Environment Institute, Climate Analytics, E3G, International Institute for Sustainable Development and United Nations Environment Programme. https://doi.org/10.51414/sei2023.050. Accessed 25 July 2024.

Song, Q., Li, J., and Zeng, X. (2015) Minimizing the Increased Solid Waste Through Zero Waste Strategy. *Journal of Cleaner Production, 104*: 199–210.

Stedman Jones, D. (2014). *Masters of the Universe: Hayek, Friedman, and the Birth of Neoliberal Politics.* Princeton University Press.

Stern, N. (2007). *The Economics of Climate Change.* Cambridge University Press.

Stigliz, J. (2024). Freedom for the Wolves, *The Atlantic*, 22 April. www.theatlantic.com/ideas/archive/2024/04/neoliberalism-freedom-markets-hayek/678124/. Accessed 24 July 2023.

Supran, G. and Oreskes, N. (2020). Addendum to 'Assessing ExxonMobil's Climate Change Communications (1977–2014)', *Environmental Research Letters*, *15*: 119401.

Supran, G. and Oreskes, N. (2017). Assessing ExxonMobil's Climate Change Communications (1977–2014), *Environmental Research Letters*, *12*: 084019.

Supran, G., Rahmstorf, S., and Oreskes, N. (2023). Assessing ExxonMobil's Global Warming Projections, *Science, 379 (6628)*. www.science.org/doi/10.1126/science.abk0063 Accessed 25 July 2024.

Swarup, V. (2020). Comment on 'Assessing ExxonMobil's Climate Change Communications (1977–2014)', *Environmental Research Letters*, *15*: 118001.

UNEA UNEP (United Nations Environment Assembly of the United Nations Environment Programme). (2022). *End Plastic Pollution: Towards an International Legally Binding Instrument.* https://wedocs.unep.org/bitstream/handle/20.500.11822/38522/k2200647_-_unep-ea-5-l-23-rev-1_-_advance.pdf?sequence=1&isAllowed=y.4. Accessed 19 October 2024.

United Nations. (n.d.). *Climate Action.* www.un.org/en/climatechange/cience/causes-effects-climate-change#:~:text=Fossil%20fuels%20–%20coal%2C%20oil%20and,they%20trap%20the%20sun's%20heat. Accessed 16 June 2024.

United Nations. (2015). *Sustainable Development Goals.* Department of Economic and Social Affairs. https://sdgs.un.org/goals. Accessed 15 June 2024.

Urry, J. (2010). Consuming the Planet to Excess, *Theory, Culture and Society*, *27(2–3)*: 191–212.

US Department of Agriculture Economic Research Service. (2009). *Access to Affordable and Nutritious Food: Measuring and Understanding Food Deserts and Their Consequences.* www.ers.usda.gov/webdocs/publications/42711/12716_ap036_1_.pdf?v=41055. Accessed 19 October 2024.

US Department of Energy. (n.d.). *Batteries for Electric Vehicles.* https://afdc.energy.gov/vehicles/electric-batteries. Accessed 18 June 2024.

US Environmental Protection Agency. (2024). *Resources, Waste and Climate Change.* www.epa.gov/smm/resources-waste-and-climate-change. Accessed 17 June 2024.

Veblen, T. (1912). *The Theory of the Leisure Class.* Macmillan.

Willsher, K. (2023). Stitch in Time: France to Help Pay for Clothes to be Mended to Cut Waste. *The Guardian*, 12 July. www.theguardian.com/environment/2023/jul/12/stitch-in-time-france-to-help-pay-for-clothes-to-be-mended-to-cut-waste. Accessed 28 July 2023.

Yu, E.P., Luu, B.V., and Chen, C. (2020) Greenwashing in Environmental, Social and Governance Disclosures. *Research in International Business and Finance*, *52*: 101192.

INDEX